Cambridge IGCSE®
Biology

Practical Teacher's Guide

Matthew Broderick

CAMBRIDGE
UNIVERSITY PRESS

University Printing House, Cambridge CB2 8BS, United Kingdom

One Liberty Plaza, 20th Floor, New York, NY 10006, USA

477 Williamstown Road, Port Melbourne, VIC 3207, Australia

4843/24, 2nd Floor, Ansari Road, Daryaganj, Delhi – 110002, India

79 Anson Road, 06–04/06, Singapore 079906

Cambridge University Press is part of the University of Cambridge.

www.cambridge.org

It furthers the University's mission by disseminating knowledge in the pursuit of education, learning and research at the highest international levels of excellence.

Information on this title: www.cambridge.org/9781316611050

© Cambridge University Press 2017

This publication is in copyright. Subject to statutory exception and to the provisions of relevant collective licensing agreements, no reproduction of any part may take place without the written permission of Cambridge University Press.

First published 2017

20 19 18 17 16 15 14 13 12 11 10 9 8 7 6 5 4 3 2 1

Printed in the United Kingdom by CPI Group (UK) Ltd, Croydon CR0 4YY

A catalogue record for this publication is available from the British Library

ISBN 978-1-316-61105-0 Paperback

The questions, answers and annotation in this title were written by the author and have not been produced by Cambridge International Examinations.

Cambridge University Press has no responsibility for the persistence or accuracy of URLs for external or third-party internet websites referred to in this publication, and does not guarantee that any content on such websites is, or will remain, accurate or appropriate. Information regarding prices, travel timetables, and other factual information given in this work is correct at the time of first printing but Cambridge University Press does not guarantee the accuracy of such information thereafter.

Every effort has been made to trace the owners of copyright material included in this book. The publishers would be grateful for any omissions brought to their notice for acknowledgement in future editions of the book.

The questions, answers and annotation in this title were written by the author and have not been produced by Cambridge International Examinations.

....

NOTICE TO TEACHERS IN THE UK

It is illegal to reproduce any part of this work in material form (including photocopying and electronic storage) except under the following circumstances:
(i) where you are abiding by a licence granted to your school or institution by the Copyright Licensing Agency;
(ii) where no such licence exists, or where you wish to exceed the terms of a licence, and you have gained the written permission of Cambridge University Press;
(iii) where you are allowed to reproduce without permission under the provisions of Chapter 3 of the Copyright, Designs and Patents Act 1988, which covers, for example, the reproduction of short passages within certain types of educational anthology and reproduction for the purposes of setting examination questions.

....

IGCSE ® is the registered trademark of Cambridge International Examinations

Contents

Introduction		v
Safety section		vi

1	**Classification**	1
	1.1 Drawing and labelling organisms	1
	1.2 Observation and drawing of pollen tubes	3

2	**Cells**	5
	2.1 Observing plant cells	5
	2.2 Observing animal cells	6
	2.3 Drawing different specimens	7
	2.4 Measuring and calculating the size of specimens	8

3	**Movement in and out of cells**	11
	3.1 Diffusion in gelatine products	11
	3.2 Osmosis in potatoes	12
	3.3 Osmotic turgor	13
	3.4 Planning an investigation into osmosis	14

4	**The chemicals of life**	17
	4.1 Testing for the presence of carbohydrates	17
	4.2 Extracting DNA	18
	4.3 Testing foods	19

5	**Enzymes**	21
	5.1 Effect of amylase on starch	21
	5.2 Effect of temperature on enzyme activity	22
	5.3 Effect of pH on enzyme activity	23

6	**Plant nutrition**	26
	6.1 Testing leaves for the presence of starch	26
	6.2 Light as a limiting factor for photosynthesis	27
	6.3 Effect of light intensity on oxygen production in Canadian pondweed	28

7	**Animal nutrition**	30
	7.1 Measuring the energy content of foodstuffs, part I	30
	7.2 Measuring the energy content of foodstuffs, part II	31
	7.3 Mouthwash versus acids	32

8	**Transport in plants**	34
	8.1 Transport of water through plants via xylem	34
	8.2 Testing the product of transpiration	35
	8.3 How environmental factors affect the rate of transpiration	36

9	**Transport in animals**	38
	9.1 Dissecting a heart	38
	9.2 Effect of exercise on heart rate, part 1	39
	9.3 Effect of exercise on heart rate, part II	40

10	**Pathogens and immunity**	42
	10.1 Culturing bacteria	42
	10.2 Bacteria around you	43
	10.3 Effect of antibacterial mouthwashes on bacteria	45

11	**Respiration and gas exchange**	47
	11.1 Germinating peas	47
	11.2 Lung dissection	48
	11.3 Effect of exercise on breathing rate	49
	11.4 Repaying the oxygen debt	50

12	**Excretion**	52
	12.1 Kidney dissection	52
	12.2 Expired and inspired air	53

13	**Coordination and response**	55
	13.1 Measuring reaction times	55
	13.2 Sensitivity test	56
	13.3 Human responses	56

14	**Homeostasis**	58
	14.1 Controlling body temperature	58
	14.2 Effect of body size on cooling rate	59
	14.3 Evaporation rates from the skin	60

15	**Drugs**	**62**
	15.1 Effect of caffeine on reaction times	62
	15.2 Effect of antibiotics on bacteria	63

16	**Reproduction in plants**	**66**
	16.1 Structure of a flower	66
	16.2 Oxygen for germination	67
	16.3 Measuring the effect of temperature on the germination of cress seeds	68

17	**Reproduction in humans**	**70**
	17.1 Protecting the fetus	70

18	**Inheritance**	**72**
	18.1 Cloning a cauliflower	72

19	**Variation and natural selection**	**74**
	19.1 Variation in humans	74
	19.2 Adaptive features	75

20	**Organisms and their environment**	**77**
	20.1 Using a quadrat	77
	20.2 Making compost	78

21	**Biotechnology**	**80**
	21.1 Effect of pectinase on apple pulp	80
	21.2 Effect of temperature on pectinase	81
	21.3 Biological washing powders	82

22	**Humans and the environment**	**84**
	22.1 Effect of acid on the germination of cress seeds	84
	22.2 Fossil fuel combustion	85

Introduction

It is not uncommon for science teachers to lack confidence in hands-on practical work. Some experiments and investigations are difficult to set up and, if they go wrong, they can waste valuable lesson time. It is frequently easier simply to teach as if science was a theoretical subject. Having said that, science is and always has been a practical subject. The skills needed for students to become successful scientists can, of course, be taught from a book or computer, but the chances of their developing a deep understanding and passion for the subject are much reduced by this method of teaching.

The aim of the Practical Workbook, Teacher's Guide and the content of the CD-ROM is to give you, the teacher, confidence to attempt a wide variety of practical work with your students. Many of the investigations are relatively simple and require little specialist equipment, and alternatives are also suggested where possible. Limitations of both equipment and time often place pressure on a teacher's ability to deliver practical work, and it is understood that, even with the best of intentions, it may not be possible to complete every investigation in this book successfully. To enable students to develop skills even if they are unable to complete all the practical work described here, the CD-ROM contains sample data that can be shared with students. This can be used either when the practical cannot be completed or when the experiment goes awry and does not provide useful data.

Students learn most by doing and being active so, although demonstrations are suggested at times to allow for the differences in student ability, getting the students to complete the investigations for themselves is strongly recommended.

Each investigation includes questions to encourage students to think about what their results mean and how they can improve the methods they have used. Getting the students to work like scientists will greatly improve their ability to think like scientists.

There is a safety section which outlines some basic safety precautions at the beginning of the book, and specific safety issues are highlighted for each investigation. This does not remove the teacher's responsibility for ensuring that all work attempted by the students is safe, and a degree of judgement will be required as to individual students' ability to follow instructions and work safely.

Safety section

Despite using Bunsen burners and chemicals on a regular basis, the science laboratory is one of the safest classrooms in a school. This is due to the emphasis on safety and the following of precautions set out by regular risk assessments and procedures.

Responsibility for all safety matters rests with the teacher and their technicians. The safety precautions set out in this guide are guidance towards your own full risk assessment that you must carry out and record.

There are different standards and guidelines in different educational authorities; it is advisable that you meet the minimum standards as set out by your local authority or educational provider. Use the following sources for further guidance as part of your laboratory risk assessment procedures.

- CLEAPSS provides support in practical science.
- COSHH offers further guidance relative to UK law and regulations. These are an excellent guide to follow if you wish to carry out a thorough risk assessment for your laboratory or department.
- MSDS (Material Safety Data Sheets) provide detailed information about the different substances and materials that you may use. This will include information such as safe storage, hazard rating, expiration times, fire hazards, preventative action, emergency action, and much more about any substance. These should be provided by your technician with the materials or solutions every time they are delivered to your classroom.
- Hazcards also offer the important safety, storage and emergency action information for chemicals used in the science laboratory.

The safety information provided in the student workbook are the basic precautions to be followed in the laboratory. The information is also designed to help the students develop the necessary planning skills to help to prepare for examination in this area. The safety precautions provided in this teacher's guide are to guide you in delivering safe science investigations in your laboratory.

The new hazard codes will be used where relevant and in accordance with information provided by CLEAPSS.

C	= corrosive		MH	= moderate hazard
HH	= health hazard		T	= acutely toxic
F	= flammable		O	= oxidising
N	= hazardous to the aquatic environment			

1 Classification

> **Overview**
>
> In this chapter, students will review the main characteristics of different organisms and identify different groups of organisms using their features. They should also be able to identify the key features of a living organism in a living thing.

Practical investigation 1.1
Drawing and labelling organisms

Planning the investigation

In this investigation, students should use their observation skills to identify possible samples for collection. These samples should then be drawn and labelled. Details of observable features should be made with reference to the guidelines for making accurate scientific drawings. The investigation can last for up to an hour, depending on the amount of time it takes to collect specimens from outside.

This investigation will focus on the following assessment objectives:
- AO3.1 Demonstrate knowledge of how to safely use techniques, apparatus and materials (including following a sequence of instructions where appropriate)
- AO3.3 Make and record observations, measurements and estimates.

Setting up for the investigation

Equipment per group of 2–4 students: small tray or box, forceps (or tweezers/small shovel for picking up items), latex gloves, sharp pencil, insect pooter.

You will need to review your own local environment prior to this investigation. Which areas on your school grounds might have suitable organisms for collection? These can be plants, grasses and flowers or small invertebrates if you have access to them. If you do not have access to suitable organisms, purchase plastic models for students to draw.

Organisms can be collected using an insect pooter – you can make your own if you do not have access to such equipment. To do so, you will need a glass jar with a lid, scissors, plasticine®, muslin, a small elastic band and some plastic tubing (approximately 7–10 mm in diameter; a large straw would suffice). You need to cut the plastic tubing into two lengths of between 15 and 25 cm, make two holes in the jar lid with scissors and insert the tubing into each hole so that at least 2–3 cm is inside the jar. Use the Plasticine® to fill any gaps in the lid and place a piece of muslin (or similar) over the end of one of the straws, using the elastic band to tie it in place. Put the lid on the jar and you now have a working pooter (see Figure 1.1). To use the pooter, you must place the end of a tube near a small insect and suck through the other tube. The air will be drawn into the jar and the insect with it. You can complete this investigation without a pooter and focus on collecting plant life if your environment dictates this. Students can use gloves and any collection equipment such as forceps, tweezers, or small shovels to collect different items from the field, playground, or other outdoor areas that are available.

Figure 1.1

Safety considerations

Tell students exactly where on the school grounds they can and cannot search for organisms. You may also need to be aware of various organisms that may be harmful in some environments, for example stinging nettles, snakes or insects that may bite. Students should wash their hands at the end of the investigation.

Common errors to be aware of

Students may have misconceptions from prior learning; they may have used the term 'micro-organisms' when talking about small organisms or invertebrates. It is best to draw this out and clarify early on. Students using the pooter will be naturally apprehensive and may not suck hard enough. A good demonstration should allay this. Students will often use shading and colouring if this is their first attempt at biological drawings; allow them to make this mistake and then learn from the self/peer assessment activity to reinforce the expectations of an accurate drawing.

Supporting your students

Some students (and teachers) will proclaim to have terrible art skills or lack confidence in drawing what they see. Demonstrate how smooth lines for the outline of the specimen and the main features give a suitable shape and proportion to what they are drawing.

Encourage light use of a sharp pencil to allow mistakes to be easily erased and rectified.

Give the more able students the responsibility of identifying the names of organisms and any key features that they can see. Ask students to use these features to classify or place the organism into groups that they know of, such as vertebrates and invertebrates. If they complete identification their own organisms, they can help other students identify theirs.

Discussion points and scientific explanation

Student drawings should aim to meet the criteria set out in the student workbook. This will guide students towards the standards required for a biological drawing. Select different drawings to highlight the difference between those who have met all of the drawing criteria and those that have not. Ask students to point out how to improve drawings. Use the discussion session to reinforce or build upon prior knowledge of the different groups of organisms that you have been working on in class. This may be the different types of plants, different vertebrates and invertebrates, or even the binomial names of the organisms found.

Answers to workbook questions

1. Student drawings
2. Handling data: formula (magnification = $\frac{\text{image size}}{\text{actual size}}$) should be used and x used to denote magnification. Answer should be above 1 to show that the drawing is larger than the actual specimen.
3. Student marks for drawings
4. Student evaluation of areas for improving drawings
5. Evaluation: table should include sensitivity and reproduction as well as any suitable features that are observable on the specimens drawn by the student.

Practical investigation 1.2 Observation and drawing of pollen tubes

Planning the investigation

You should make an advance purchase of a quantity, and range, of different flowers that are medium-large, easy to dissect and contain pollen tubes that are highly visible. Daffodils are a good example but you may need to find something similar in your local environment or flower shop. It may be that you give the whole class the same type of flower but a range will encourage them to compare the differences between them and/or make more than one type of drawing. Scalpels or suitable, sharp knife is required as well as an area for dissection. This may be a dissection tray, chopping boards or on the workbench itself. The investigation should take no longer than 20 minutes to dissect and make suitable drawings. This investigation will focus on the following assessment objectives:

- AO3.1 Demonstrate knowledge of how to safely use techniques, apparatus and materials (including following a sequence of instructions where appropriate)
- AO3.3 Make and record observations, measurements and estimates.

Setting up for the investigation

Equipment per group of 2–4 students: scalpel of knife, dissection tray, or board, different flowers with pollen tubes as described above.

Give each student at least one flower to allow each student a cross-section to draw. Students can make more than one drawing on other paper if required and groups could share flowers if you have purchased more than one type of flower.

Safety considerations

Demonstration of safe use of the scalpel or knife is important. Students may need to carry the scalpel across the classroom and a safe method of doing this should be demonstrated, carrying the scalpel with the blade not pointing outwards, walking slowly, and being aware of the surroundings and other people.

Common errors to be aware of

Students may not grip the flower correctly or may not make a clean cross-section to show the pollen tubes. Demonstrate clearly before allowing students to begin.

Supporting your students

If students are struggling with the dissection skills, either the teacher can do this in advance or do so and take photographs of the cross-sections of the flower. This will allow more time to focus on the drawing skills and the labelling of the parts of the flower and the pollen tubes.

Challenging your students

Students can use books and/or the internet to label fully their cross-sections.

Discussion points and scientific explanation

Bring in the knowledge and description of the different angiosperms; monocots and dicots. Link to the different shapes lf leaves, for example, that these classes of plants might be expected to exhibit. Discussion of the labelled parts can be related to being able to describe what they see when faced with questions or activities like this one.

Answers to workbook questions

1. Any of the criteria that the student has met in their drawing
2. Any of the criteria that the student has not met in their drawing
3. Answer depends on the flower used

Answers to exam-style question

1. **a** Award marks as follows for the student drawing:
   - smooth outline drawn with a sharp pencil [1]
   - shape and proportion appropriate [1]
   - all observable features of the specimen drawn correctly [1]
   - all features labelled such as wings, body parts, number of legs, antennae [1]
   - drawing is larger than the original picture [1].

   b Student answer ± 1 mm [1]

   c Student answer ± 1 mm [1]

   d Correct formula used [1], correct numbers from previous answers used [1], answer is greater than 1 with correct units. [1]

   e The yellow-fever mosquito is from the genus *Aedes*. [1]

   f Mosquitos are insects and belong in the Class Insecta. [1]

   Total [12]

2 Cells

Overview

In this chapter, students will review the different structures and organelles that make up different cells in the plant and animal kingdoms. They will observe the similarities and differences between the levels of organisation and be able to recognise these using a light microscope. They will also learn how to calculate the size of specimens and have further opportunities to practise the skill of producing a biological drawing of a specimen learnt in Chapter 1.

Practical investigation 2.1
Observing plant cells

Planning the investigation

In this investigation, students will develop their microscope, observation and biological drawing skills. These require repeated practice and this chapter presents the ideal opportunity to master these skills before moving on. Students will also plan a safe investigation and use appropriate formulae to carry out calculations related to what is viewed down a microscope. The investigation will typically take 30–40 minutes to complete. This investigation will focus on the following assessment objectives:

- AO3.1 Demonstrate knowledge of how to safely use techniques, apparatus and materials (including following a sequence of instructions where appropriate)
- AO3.3 Make and record observations, measurements and estimates
- AO3.5 Evaluate methods and suggest possible improvements.

Setting up for the investigation

Equipment list per group (2–3 students): small piece of (red) onion (one onion will provide enough samples for 10–40 students), staining solution (1% methylene blue or iodine), scalpel, forceps, mounted needle (the end of a pencil can be used), microscope slide, cover slip, light microscope, safety spectacles, disposable gloves, filter paper or paper towels.

Equipment can be made available in one area of the room so that students, using their equipment list, can identify and select the appropriate equipment. Demonstrate how to carry the microscope safely, with one hand clasping the neck and one hand firmly on the base.

Safety considerations

MH

- Wear safety spectacles at all times.
- Care taken when using and storing the scalpel or knife.
- The microscope lens can become very hot. Handle with care and keep turned off when not in use.
- Methylene blue and iodine are irritants. Wear gloves to protect, wash hands immediately if in contact with solution, and refer to safety information for the chemical before use.

Common errors to be aware of

Most errors can be avoided by demonstrating the procedure for preparing a slide and using a microscope. Some students may be using a microscope or preparing a specimen for the first time. The investigation could be broken up into smaller stages: set up the equipment; find a sample; prepare the slide; use of microscope. Students may benefit from seeing examples of what they are looking for. This can be done with a video microscope or by using a smartphone or camera to take a photograph through the eyepiece lens. Alternatively, display an image found on the internet on the board.

Supporting your students

Some students will struggle focusing the microscope to view the specimen. If breaking the method down into smaller stages does not help, then a one-to-one demonstration of the technique required to turn the focusing wheel will help. You can stress the need for a slow, steady hand. Some students will benefit from the teacher finding the specimen for them and then moving it slightly out of focus. This instills confidence in the student that the correct focus is not far away.

Challenging your students

This is often an opportunity for all of the students to demonstrate their excellence, and not just the usual two or three students who regularly score the highest in tests and assessments. Some students with excellent motor skills may find, and draw, their specimens with ease. These students can take on a 'teacher' role and support/show other students how to focus on their specimen. By talking through the steps with other students, these students will be talking through the steps required for investigation planning.

Discussion points and scientific explanation

Display an image of the cells on the board and ask students to describe both the observed structures and those that cannot be observed. The light microscope works by allowing light to pass through the cells. The light microscope will increase the image size of the specimen to allow you to observe the image in detail. The light microscope will only allow you to observe the cell wall, nucleus, cytoplasm, vacuoles and, in special circumstances, the location of the cell membrane. An electron microscope would be required to observe smaller structures such as the cell membrane and mitochondria.

Answers to workbook questions

1. Students should now be able to assess their own drawings as shown in Chapter 1.
2. The total magnification should be the appropriate multiplication of the two lenses in use. The units used should include a × to indicate magnification.
3. Nucleus, cell wall, cytoplasm
4. Cell membrane, vacuole, ribosomes, any other appropriate cell structure except chloroplast as these are not found in onion cells
5. In order to see the cells clearly, i.e. so that layers of cells did not overlap each other
6. To see the cells more clearly

Practical investigation 2.2
Observing animal cells

Planning the investigation

In this investigation, students will build on their observation and drawing skills from Practical investigation 2.1. They will have more independence and they should require less guidance from you in preparing their slides and assessing their drawings. This investigation will focus on the following assessment objectives:

- AO3.1 Demonstrate knowledge of how to safely use techniques, apparatus and materials (including following a sequence of instructions where appropriate)
- AO3.3 Make and record observations, measurements and estimates
- AO3.5 Evaluate methods and suggest possible improvements.

Setting up for the investigation

Equipment list for groups of 2–3 students: light microscope, cotton bud, disinfectant solution (Vircon or ethanol solution can also be used), safety spectacles, disposable gloves, staining solution (iodine or methylene blue are ideal), microscope slide, cover slip, mounted needle.

Students will use their prior knowledge in most parts of this investigation but will need a clear demonstration of the technique for ascertaining and applying the cheek cells. The cotton bud should be rotated with some force on the inside of the cheek and then the sample applied in the same manner to the centre of the microscope slide. Remove excess staining solution using filter paper or a paper towel. The investigation should take no more than 20–30 minutes depending on the microscope ability of the group.

Safety considerations
MH   HH

- Wear safety spectacles and gloves at all times.
- Place cotton buds into the disinfectant solution immediately after use.
- The microscope lens can become very hot. Handle with care and keep turned off when not in use.
- Methylene blue and iodine are irritants. Wear gloves to protect, wash hands immediately if in contact with solution, and refer to safety information for the chemical before use.

Common errors to be aware of

Students may not recall where they swabbed their cheek cells on the slide. This can be easily avoided by directing them to swab as close to the centre of the slide as possible. Students will often find air bubbles and mistake them for animal cells. If this happens, have a photo, or your own example, ready for students to see what they should be looking for.

Supporting your students

Have samples or pictures of animal cells under the microscope ready for students to know what they are looking for. Allow them to try to find the cells without help at first, but they are certainly helpful to have to hand.

Challenging your students

Ask students to help other students as they may have done during Practical investigation 1.1. This time, ask them to produce a mini guide for those who struggled – this guide should describe and explain the different stages, with diagrams to support. This could be a series of instructions and/or diagrams to help the less able students through the stages of finding and drawing a biological specimen.

Discussion points and scientific explanation

Discuss the different organelles that can and cannot be observed under the microscope. Ask them what sort of microscope would allow them to observe other organelles and which organelles they would be able to view.

Answers to workbook questions

1. Student methods should include a sensible systematic method, name all relevant equipment, and explain clearly how to prepare the microscope slide for viewing.
2. Any five from the following; smooth outlines, drawn with a sharp pencil, specimen drawn in the appropriate shape and proportion, drawing is large and clear, observable features labelled, labels drawn with a ruler, no shading or colours used
3. Diagrams should meet the criteria in the answer for Question 2.
4. Nucleus in both cells, cell wall in plant cell, allow cell membrane for animal cell, allow cytoplasm for both cells
5. Mitochondria
6. Answers should refer to a microscope of higher magnification such as an electron-scanning microscope.
7. Answers should refer to destroying bacteria or preventing bacteria from spreading.

Practical investigation 2.3
Drawing different specimens

Planning the investigation

In this investigation, students will focus on their microscope skills to explore other types of cells and tissues that you may have available in your science department. This is an opportunity to find different specimens under the microscope and to draw them. The aim is to score maximum marks for the biological drawing, something that should be easier to do after two previous attempts and feedback. The investigation can be as quick as 10 minutes but, depending on the number of slides that you have available, it can stretch to over 30 minutes.

This investigation will focus on the following assessment objectives:
- AO3.1 Demonstrate knowledge of how to safely use techniques, apparatus and materials (including following a sequence of instructions where appropriate)
- AO3.3 Make and record observations, measurements and estimates.

Setting up for the investigation

Equipment list for groups of 2–4 students: light microscope, pre-mounted slides of different cells and tissue. It does not matter how obscure or common these slides are; the aim is to focus clearly on the specimen and accurately draw what is seen. Some examples include: ciliated cells, root hair cells, body parts of different invertebrates, xylem vessels, palisade cells, blood cells. Anything pre-mounted onto a microscope slide which can be seen clearly is suitable for this activity.

Safety considerations

Broken glass or slides to be reported to the teacher.

Supporting your students

Allow struggling students to use the slides with the most detail, for example a large piece of tissue or an insect leg. Avoid giving these students a slide that contains one single cell that is not visible to the naked eye.

Challenging your students

Slides will usually have the binomial or descriptive name of the specimen on it. Allow students access to the internet or reference books to get more information on what they are looking at. Allow students to explain to the rest of the class what they found out as a mini plenary towards the end of the lesson.

Discussion points and scientific explanation

Ask students to name and describe the different levels of organisation that they observed under the microscope. Ask students to identify the different cells and tissue that they observed and, if possible, see if students can make links between them.

Answers to workbook questions

1. Students should be able to meet the following criteria: smooth outlines, drawn with a sharp pencil, specimen drawn in the appropriate shape and proportion, drawing is large and clear, observable features labelled, labels drawn with a ruler, no shading or colours used.
2. A collection or group of cells of similar shape and size that perform the same function. Allow extra credit for a named example such as muscle tissue, nerve tissue or skin tissue.
3. The named organ relates to the named tissue from Question 2. An example would be that skin contains skin tissue.
4. Student answer should name the focusing wheel and the fine focusing wheel with some description of moving them slowly through the different magnifications until they observe the specimen as clearly as possible.

Practical investigation 2.4
Measuring and calculating the size of specimens

Planning the investigation

In this investigation, students will continue to develop their microscope skills and plan a suitable method and the equipment needed for observing animal (cheek) cells. This is a similar investigation to Practical investigation 2.2 but it is important that students have the opportunity to master the skills of using a microscope. This is one of the earliest opportunities in the syllabus to explore mathematics, standard units, and sub-levels of standard units. This investigation aims to bring together all of the skills from the first two chapters as well as introducing mathematical skills. Set the planning phase as homework or do in another lesson. This investigation will focus on the following assessment objectives:

- AO3.1 Demonstrate knowledge of how to safely use techniques, apparatus and materials (including following a sequence of instructions where appropriate)

- AO3.2 Plan experiments and investigations
- AO3.3 Make and record observations, measurements and estimates
- AO3.4 Interpret and evaluate experimental observations and data.

Setting up for the investigation

Equipment list for groups of 2–3 students: light microscope, cotton bud, disinfectant solution (Vircon or ethanol solution can also be used), safety spectacles, disposable gloves, staining solution (iodine or methylene blue are ideal), microscope slide, cover slip, mounted needle.

Students should plan a similar method to the method used in Practical investigation 2.2. Give students the option of writing the method and beginning immediately, or have them check the method with you first. The investigation should take between 20 and 40 minutes depending on the group sizes and planning time allowed.

Safety considerations

MH

- Wear safety spectacles and gloves at all times.
- Place cotton buds immediately into the disinfectant solution before and after use.
- The microscope lens can become very hot. Handle with care and keep turned off when not in use.
- Methylene blue and iodine are irritants. Wear gloves to protect, wash hands immediately if in contact with solution, and refer to safety information for the chemical before use

Supporting your students

Students who are struggling with the planning should be directed back to Practical investigation 2.2 to view the original method. Measuring the specimen is often difficult so you may choose to do this as a Stretch and challenge activity for the more able students. In this case, you may provide the less able students with examples for calculating the magnification or actual specimen size.

Challenging your students

The more able students can produce a table for the size calculations. Ask them to produce a table with headings and units that would allow a less able student to enter their results systematically. The purpose of the table should be to make it a step-by-step process for entering the data and calculating the specimen size.

Discussion points and scientific explanation

Class discussion about how to approach the planning of an investigation, making sure that the equipment, method and safety rules are covered thoroughly. Review the answers given for the determined sizes of the actual specimen and compare this to prior knowledge of how big an animal cell should be.

Answers to workbook questions

1. All relevant equipment should be listed for the method given in **2**.
2. Answer should describe the preparation of a cheek cell specimen.
3. Safety precautions should mention the safe use of the staining solution, the wearing of safety spectacles and gloves, and the disposal of the cotton bud into a disinfectant solution.
4. Students should be able to meet the following criteria: smooth outlines, drawn with a sharp pencil, specimen drawn in the appropriate shape and proportion, drawing is large and clear, observable features labelled, labels drawn with a ruler, no shading or colours used.
5. Student calculation
6. Student should refer to the uniform nature of the onion cells as they have a regular shape and may stretch across the field of view.
7. Any named plant cell such as palisade cell, mesophyll cell, guard cell

Answers to exam-style questions

1.
   a. The drawing should be larger than the original [1], smooth lines drawn with a pencil [1], shape and proportion of major features correct [1], suitable labels of main features such as shell, eye, foot, tentacles, unsegmented body [2].
   b. length ± 1 mm [1]
   c. length ± 1 mm [1]
   d. correct formula used [1], lengths correctly entered [1], answer is greater than 1 with units [1]

2.
   a. Wearing safety spectacles and gloves [1], disposal of cotton buds [1], using disinfectant or similar solution [1]
   b. Nucleus, cytoplasm [2]
   c. Actual size = image size ÷ magnification [1], actual size = $0.05\,\mu m$

3 Movement in and out of cells

> **Overview**
>
> In this chapter, students will observe the processes of diffusion and osmosis. This provides them with the opportunity to test the predictions that they make about how some substances will behave when placed in different solutions.

Practical investigation 3.1 Diffusion in gelatine products

Planning the investigation

In this investigation, students will observe a short investigation into diffusion. This will support the understanding of diffusion and the movement of molecules from an area of higher concentration to an area of lower concentration. Students will also have an opportunity to suggest possible improvements to the investigation.

This investigation will focus on the following assessment objectives:

- AO3.1 Demonstrate knowledge of how to safely use techniques, apparatus and materials (including following a sequence of instructions where appropriate)
- AO3.2 Plan experiments and investigations
- AO3.3 Make and record observations, measurements and estimates
- AO3.4 Interpret and evaluate experimental observations and data
- AO3.5 Evaluate methods and suggest possible improvements.

Setting up for the investigation

Equipment list (per student or pair of students): red gelatine (should contain red creosol), scalpel or knife, 1M hydrochloric acid, test-tube and bung, safety spectacles.

Some gelatine contains meat products but non-meat alternatives will be available in your local supermarket. If gelatine is difficult to get hold of, this investigation can be replicated using Activity 3.3 from the coursebook.

This is a short investigation and the only preparation required is the advance purchase of gelatine that contains red creosol. Red creosol solution can also be bought from your local chemical suppliers. The investigation should take no longer than 15 minutes to set up and carry out.

Safety considerations

C

- Take care when using the scalpel.
- Wear safety spectacles at all times.
- Consult safety data for use and storage of hydrochloric acid.

Supporting your students

Draw a diagram to represent the differing concentrations inside and outside the gelatine. Do this on paper or on the board if many students require help. Once the students can see some sort of concentration gradient, they will be able to see, and describe, what is happening.

Challenging your students

Ask students to cut pieces of gelatine in different sizes to observe how surface area-to-volume ratios affect the rate of diffusion. Students to try at least three different sizes and record the time taken for the gelatine to turn from red to yellow.

Discussion points and scientific explanation

Students discuss how the acid diffuses into the gelatine and that the evidence for this is the change in colour from red to yellow of the gelatine. Students should be able to talk about the concentration gradient or the net movement of the molecules from higher concentration to lower concentration.

> **Answers to workbook questions**
>
> 1. Gelatine turns to yellow in 1–minute, depending on the concentration of HCl used.
> 2. Student answer explains that diffusion has occurred and some of the acid solution has moved from an area of higher concentration to an area of lower concentration (inside the gelatine). When this happens, the gelatine turns from red to yellow as this is what happens to the chemical (red creosol) inside gelatine in acidic conditions.
> 3. Student should suggest control variables and how to keep them the same. The size of the gelatine block, the amount of acid, the pH of the acid, the temperature of the room, using the same equipment are all acceptable answers.

Practical investigation 3.2
Osmosis in potatoes

Planning the investigation

In this investigation, students will compare how osmosis happens in different solutions. They will practise their skills of making observations and collecting data. They will explain the movement of water in or out of the potato tissue samples using their knowledge of osmosis. This investigation will focus on the following assessment objectives:

- AO3.1 Demonstrate knowledge of how to safely use techniques, apparatus and materials (including following a sequence of instructions where appropriate)
- AO3.2 Plan experiments and investigations
- AO3.3 Make and record observations, measurements and estimates
- AO3.4 Interpret and evaluate experimental observations and data.

Setting up for the investigation

Equipment list per group of 2–4 students: medium-sized potato, cork borer (a knife can be used if not available), plastic ruler, distilled water, test-tubes ×6, scalpel or knife, marker pen for writing on test-tubes, safety spectacles, sucrose solution. It is recommended that you use 30% and 70% sucrose solution but any sugar solution of differing concentrations can be used. The column for solutions is blank in the student table to allow you to choose the solutions of your choice. The investigation will take between 30 and 45 minutes to complete.

As long as you have distilled water and some sort of sugar solution, this investigation will work well. Extend to include more concentrations of the sugar solution if you wish, or it can be a simple comparison between water and sugar solution. The longer that you leave the potato tissue samples in the solution, the more pronounced the results will be. To save time with very large classes, it may be helpful to get the technician to prepare some, or all, of the potato samples.

Safety considerations
MH

- Take care using the cork borer and scalpel.
- Wear safety spectacles at all times.

Common errors to be aware of

Pieces of potato get stuck inside the cork borer but these can be easily removed by poking through with a thin implement or pencil. Some students may not mark their test-tubes effectively or get them mixed up so a clear demonstration is key to ensure that good practice is followed.

Supporting your students

Assigning roles within small groups will provide support for struggling students. Less able students can focus on the planning skills by collecting the equipment required. Identify some students to prepare the samples and organize the test-tubes, and other students to complete the table and enter the data. This allows students to develop their strengths while observing their peers to develop their areas for improvement.

Challenging your students

Ask students to create an extra column to calculate the percentage change in length of the potatoes. This can be used to graph the data and shown to or discussed with the rest of the class.

Discussion points and scientific explanation

Time for discussion is important here to allow you to go through each of the results achieved in each solution. Ask students to explain on the board, using diagrams, which way water moved in or out of the potato tissue samples. Common misconceptions can be caught here, such as the idea of glucose/sugar molecules moving in and out of the potato tissue samples.

The potato tissue samples in the distilled water will increase in length as water molecules leave the higher concentration of water to enter the potato tissue samples.

The potato in the 30% sugar solution will decrease in length as the water in the potato tissue sample moves from its' higher concentration to the sucrose solution which has a lower concentration than inside the potato. The potato in the 70% sugar solution will decrease more than the potato in the 30% solution as the concentration gradient is greater and more water molecules will move out of the potato tissue samples in the sugar solution, which has a very low water concentration.

Answers to workbook questions

1. Completed student table
2. Student answer should describe how the potato strips differ in the different solutions.
3. Student answer should describe the addition of the two lengths of potato and dividing by two.
4. Describe – students should describe the changes in average length of the potatoes as increasing or decreasing, using data from the table to support their answer. Explain – students should attribute the change to osmosis. The movement of water molecules from inside or outside of the potato should be linked to the movement of water molecules from an area of higher water concentration (high water potential or low sugar concentration or dilute solution) to an area of lower water concentration (low water potential or high sugar concentration or concentrated solution).
5. Student lists as many of the control variables as possible, such as volume of solution used, surrounding conditions kept the same, same equipment used, same varieties of potato or any other sensible suggestion.
6. It is possible to pour different amounts of water, sensible suggestion such as use a measuring cylinder or similar equipment. Also, accept use mass of water to ensure the exact amounts used.

Practical investigation 3.3
Osmotic turgor

Planning the investigation

In this investigation, students will investigate osmosis using Visking tubing and a syrup solution. Students will demonstrate practical skills in setting up and carrying out the investigation, and applying their knowledge of osmosis to explain what happened. The investigation takes up to 1 hour but time can be saved with advance preparation for those with shorter lessons.

This investigation will focus on the following assessment objectives:
- AO3.1 Demonstrate knowledge of how to safely use techniques, apparatus and materials (including following a sequence of instructions where appropriate)
- AO3.3 Make and record observations, measurements and estimates.

Setting up for the investigation

Equipment list for groups of 2–4 students: safety spectacles, test-tube and rack, 20 cm Visking tubing (soaked in water in advance), distilled water, elastic band, graduated pipette (syringe can also be used), syrup solution. The syrup solution can be made by mixing golden syrup or sucrose with its own volume/weight of water.

Save time by preparing the tubing in water and tying the first knot yourself. This means students can move to adding the syrup solution straight away and the investigation can be set up within the first 5 minutes of a lesson.

Safety considerations

Wear safety spectacles during the investigation.

Common errors to be aware of

The trickiest part of this is opening up the Visking tubing. Do this by rubbing the ends between two fingers until a small opening appears. Blowing into the opening will open it further; alternatively, you can open it with the end of the pipette but be careful to avoid tearing the tubing. It is not always necessary to tie the end of the tubing to the test-tube but, if you have an elastic band, it is worth doing so to keep it fixed firmly in place. If students are using the test-tube racks, they can label their test-tubes with their name.

Supporting your students

Demonstrate what is happening by drawing the relative concentrations on the board to make it easier to see the difference in concentration inside and outside the tubing. Students will be able to point out the direction of the movement of water if they can visualise or see the relative concentrations.

Challenging your students

Ask students to plan how they could measure the change in volume of the tubing.

Discussion points and scientific explanation

The tubing reaches turgor during the investigation and is difficult to bend compared to the ease with which it bent before the investigation. Following on from the previous investigation, discuss with students how the water moved into the tubing through the semi-permeable membrane as it moved from an area of higher water concentration to an area of lower water concentration.

Answers to workbook questions

1. Student diagram to show a flaccid tubing before the investigation and a turgid tubing afterwards. Give credit for showing that the tubing could be bent before the investigation but not as much afterwards.
2. Student description should describe what they have seen or recorded in Question 1. Students are expected to use key words such as turgid in their description.
3. Student answer describes how water molecules have moved from the area of higher water concentration outside the tubing into the tubing where the water concentration is lower.
4. Two possible answers may be suggested here. The tubing may have burst as the volume and pressure increased due to the movement of water into the tubing. If the tubing was strong enough, some suggestion of equilibrium between the pressure from inside and outside of the tubing.
5. To remove any syrup from the outside of the tubing and avoid contamination of the water concentration outside of the tubing

Practical investigation 3.4 Planning an investigation into osmosis

Planning the investigation

In this investigation, students will plan their own investigation to observe and compare osmosis in different solutions. A simple equipment list has been provided to guide students towards devising a simple method to compare the mass of potato tissue samples when placed into water and a sugar solution. This investigation will focus on the following assessment objectives:

- AO3.1 Demonstrate knowledge of how to safely use techniques, apparatus and materials (including following a sequence of instructions where appropriate)
- AO3.2 Plan experiments and investigations
- AO3.3 Make and record observations, measurements and estimates
- AO3.4 Interpret and evaluate experimental observations and data
- AO3.5 Evaluate methods and suggest possible improvements.

Setting up for the investigation

Equipment list for students (this can be done individually or in groups of 2–4 depending on your class size and resources available): potato, cork borer, balance or weighing scales, safety spectacles, marker pen, distilled water, test-tubes × 4 and rack, 20% sugar solution, stopwatch. Any sugar solution can be used here.

The planning and execution of the investigation could take over an hour so consider allowing students time to plan their method either for homework or during a prior lesson. This will allow you time to review their methods and safety precautions before beginning the investigation. The investigation itself will take between 30 and 40 minutes.

Safety considerations

MH

- Safe use of the cork borer (or knife if preferred) to be highlighted.
- Clean up spillages and report broken glass to the teacher.

Common errors to be aware of

Some students may not have potato tissue samples of the same starting mass. Potato pieces may become stuck in the core borer – simply push through with the end of a pencil or similar implement.

Supporting your students

To plan collaboratively, place students into groups of similar ability. This will allow you to work with the students who may struggle with the planning aspect of this investigation. Discuss each stage with them while referring back to Practical investigation 3.2.

Challenging your students

More able students who may complete this investigation faster can improve reliability by adding a third potato sample to their method. Alternatively, students can use a third solution of their choice and prepare this in advance. Students can use the internet to research how to make a 50% sugar solution using sucrose or syrup and water.

Discussion points and scientific explanation

Discuss the main points of the investigation that students aim to cover in their planning. Did they use all of the equipment? Did students begin with the same mass of potato tissue sample at the beginning and other control variables such as the volume of solution used? Link to the need for reliability in scientific investigations.

Answers to workbook questions

1. Student method refers to preparing potato tissue samples of the same mass, using a cork borer. Student describes method of placing potato tissue samples into different solutions at the same time and recording their mass at the end of a suggested time, such as 15–20 minutes.
2. Any two sensible safety precautions such as safety spectacles and safe use of the cork borer or knife.
3. Student table to have the independent variable (type of solution) in the left-hand column and the data leading to the dependent variable (change in mass of potato tissue sample) with suitable units in the heading of the right-hand side of the table. Tables should be drawn neatly, using a pencil and ruler with sufficient space to record data.
4. Student calculates the percentage change in the mass of the potato tissue samples:

   $$\text{Percentage change} = \frac{\text{final average mass}}{\text{starting average mass}} \times 100$$

5. Student describes the change in mass using data from their table. Student explains the change in mass using osmosis to describe the movement of water in or out of the potato tissue sample from an area of higher water concentration to an area of lower water concentration.
6. Use more than two samples of potato tissue per solution.

Answers to exam-style questions

1.
   a. The starch solution turns from clear to a blue/black/purple colour.
   b. The iodine solution moved into the Visking tubing, the starch molecules were too large to move through the membrane of the tubing.
   c. Any two from safety spectacles, wearing gloves, wash hands afterwards or any other sensible suggestion

2.
   a. 0.14, 0.26, 0.13 [1], −0.11, −0.26, −0.39 [1], −0.53, −0.48, −0.59 [1], g or grams added to the heading of the table [1]
   b. Distilled water = 0.18 g [1], 20% sucrose solution = −0.25 g [1], 40% sucrose solution = −0.53 g [1]
   c. Student answer describes the movement of water [1] by osmosis [1] to increase/decrease the mass of the potato [1] in a named solution [1], accept reference to semi-permeable membrane
   d. To improve reliability of data, to spot anomalies, to take the mean

4 The chemicals of life

> **Overview**
>
> In this chapter, students will investigate how to test for the presence of carbohydrates, protein and fats in foodstuffs. This will support their knowledge of the smaller molecules that make up larger molecules, such as starch and proteins. They will also extract DNA from a sample of their own saliva.

Practical investigation 4.1 Testing for the presence of carbohydrates

Planning the investigation

In this investigation, students will carry out the positive test for reducing sugars and starch to enable them to carry out tests on real food and drinks in the future. This investigation will focus on the following assessment objectives:

- AO3.1 Demonstrate knowledge of how to safely use techniques, apparatus and materials (including following a sequence of instructions where appropriate)
- AO3.3 Make and record observations, measurements and estimates.

Setting up for the investigation

There are several components of this where alternatives may need to be used depending on your location and storage capabilities. The following list is an outline of the options; however, this is just a guide as there are many other alternatives.

Each group will need: 20% glucose solution (can be made by combining 20 g glucose/sugar with 80 ml water), vegetable oil (cooking oil can be used), alcohol (e.g. ethanol or propanol), hot water (water bath or beakers of hot/boiled water), 1% starch solution, 1% protein solution (e.g. albumen), biuret solution (5 ml dilute sodium hydroxide and 5 ml dilute copper sulfate can be used, iodine solution, Benedict's solution, pipettes, test tubes × 5, test-tube rack, water, safety spectacles.

Check the solutions for storage instructions and follow accordingly.

Safety considerations

Students should wear safety spectacles. Beware of ingestion or inhalation of substances and encourage students to report spillages to the teacher to clear up safely.

Common errors to be aware of

This investigation has many components and students will need to take their time and work methodically to avoid mixing up the many different solutions or test-tubes. Some students may not get expected results for the biuret test and a gentle shake of the test tube should resolve this. The Benedict's test needs several minutes to change colour if a Bunsen burner or heat source is not used. Usually, water boiled from a kettle will suffice and allow the colour change to take place.

Supporting your students

Break down the investigation into smaller sections – either by testing two of the substances at a time, or even just one at a time. Allowing students to observe and repeat what they see in your demonstration will support these students.

Challenging your students

Allow students to test the substances using the other solutions and reagents. This will allow them to observe the positive and the negative results for each solution. Ask students to predict what types of food they can carry out a positive test on using the solutions available in today's lesson.

Discussion points and scientific explanation

Students should be able to identify and discuss the following changes:
- Starch solution tested with iodine: iodine turns from orange/brown to blue/black/dark purple
- Protein solution tested with biuret test: turns from blue to mauve/purple
- Glucose solution (reducing sugar) tested with Benedict's: turns from blue to brick red/orange
- Vegetable oil tested with mix of alcohol and water: cloudy-white suspension/layer forms

Answers to workbook questions

1. Students should observe/record/draw what they see as outlined.
2. Starch solution tested with iodine: Student answer
3. Protein solution tested with biuret test: Student answer
4. Glucose solution (reducing sugar) tested with Benedict's: Student answer
5. Vegetable oil tested with mix of alcohol and water: Student answer

Practical investigation 4.2 Extracting DNA

Planning the investigation

In this investigation, students will carry out a short investigation to extract their own DNA. Students will then draw what they observe and link this to their knowledge of DNA structure.
This investigation will focus on the following assessment objectives:

- AO3.1 Demonstrate knowledge of how to safely use techniques, apparatus and materials (including following a sequence of instructions where appropriate)
- AO3.3 Make and record observations, measurements and estimates.

Setting up for the investigation

Equipment list (per student or pair of student): drinking water (from either taps or bottles), salt, clear drinking cups/glasses, glass beakers, washing-up liquid, food colouring, glass rods or stirrers, isopropyl alcohol (or any alcohol solution that is available, such as ethanol, methanol, etc.), safety spectacles.
Very little preparation is required here and most of the materials required should be readily available in most communities. This is a variation of extracting DNA from fruit (which could also be done) but this is exciting for students to extract and observe their own DNA.

Safety considerations

Wear safety spectacles and take care when using alcohol solution.

Common errors to be aware of

The main errors that can be encountered here include not gargling the water for long enough and the tricky task of pouring the alcohol mixture gently down the side of the saltwater cup. Demonstration of this technique should be carried out with a diagram of the appropriate pouring angle displayed on the board. The DNA should not be left for too long before observation as it will cease to be visible after a few minutes.

Supporting your students

Pair a more able student with a student who is struggling. The more able student can support the other student with the skills required to carry out the method successfully.

Challenging your students

Extract DNA from a piece of fruit using the same method and compare this with the students' own. The student can look for similarities and differences between the two samples.

Discussion points and scientific explanation

Students will want to know how the extraction of DNA is possible. Lead a discussion to explore how some cheek cells are collected by gargling the salt water. The role of the washing-up liquid is to break down the membranes of the cheek cells – this is an opportunity to relate to possible prior knowledge of enzymes. The DNA is then released into the water where it will form a solid between the alcohol and water layers. DNA is not soluble in alcohol whereas most of the other parts of the cheek cell will dissolve into the water layer. This leaves the white clumps and string that can be observed, and this is the DNA clumping together in solution.

Answers to workbook questions

1. Students should observe the DNA using words such as 'clumps' and 'stringy' or other sensible alternatives.
2. Some students will not due to incorrect pouring of the alcohol mixture or not gargling the salt water enough in order to collect enough.

Practical investigation 4.3
Testing foods

Planning the investigation

In this investigation, students will use their previous knowledge to select the appropriate materials in order to test different foods for the nutrients that they contain. This investigation will focus on the following assessment objectives:

- AO3.1 Demonstrate knowledge of how to safely use techniques, apparatus and materials (including following a sequence of instructions where appropriate)
- AO3.2 Plan experiments and investigations
- AO3.3 Make and record observations, measurements and estimates
- AO3.5 Evaluate methods and suggest possible improvements.

Setting up for the investigation

Equipment list (per student or pair of students): test tubes or spotting tiles (spotting tiles are more suitable for testing many different foods), test-tube rack, pestle and mortar to grind the food if necessary, safety spectacles, water, alcohol, pipettes, hot water bath, biuret solution (or 5 ml dilute sodium hydroxide and 5 ml dilute copper sulfate as in Practical investigation 4.1), iodine solution, Benedict's solution, range of different food and drink for students to test. Food choices should be items such as bread, cereal, pasta, biscuits, crisps, cold meat, milk, orange juice, fruits or whatever is available in your local community.

The equipment could be set out along one long workbench or table for students to be able to choose. The most important aspect of this investigation is students being able to assess what materials they have, create a plan, and then execute it successfully. Avoid creating student-friendly, pre-prepared group kits as this removes the planning aspect of the investigation.

Safety considerations

Wear safety spectacles, take care using and storing the different chemicals and be aware of potential allergies. It is better to check in advance (or consult the class medical notes) for allergies and avoid using that substance if necessary.

Common errors to be aware of

Students who rush into the practical part of the investigation may make errors so it is important to stress the need for careful planning and thought (just like you would in an assessment). Monitoring of the practical area is helpful to avoid contamination of the pipettes. The class size may dictate that this investigation is spread across two, or even three, lessons to allow students ample time and equipment to carry it out.

Supporting your students

Encourage these students to plan and execute the food tests individually rather than trying to plan too many different tests at once. Students may benefit from taking their previous notes from Practical investigation 4.1 to the planning area and work in pairs to discuss their plan of action.

Challenging your students

Students can research other foods that could have been used for each of the food groups. For example, predicting that chicken would be protein-rich is a good choice for testing for protein.

Discussion points and scientific explanation

This depends on the types of food and drink used as you can discuss which foods were rich in which nutrients. Students should be able to link their existing knowledge of the smaller molecules that make up these larger molecules (such as proteins and amino acids).

Answers to workbook questions

1. Student methods to carry out tests – Student answer
2. Two safety requirements – Student answer
3. Completed table – Student answer
4. Student answer should correctly state the colour change (from X to Y) for the named food substance.
5. Any sensible answer that suggests using the same quantities of food, repeating the investigation more than once

Answers to exam-style questions

1.
   a. Food substance A [1]
   b. Food substance C [1]
   c. Student answer should refer to the dark-red colour of the end result [1] in comparison to the weaker, orange colour of food substance B [1]
   d. Benedict's solution [1]
   e. Heat the solution in a water bath, boiling water or with a Bunsen burner.
   f. Any one sensible safety precaution such as wearing safety spectacles or gloves when handling solutions, taking care if using a knife [1]
   g. Any sensible named substance that is high in concentration for reducing sugars, such as glucose biscuit [1]

2. One mark for each of the following points:
   - equipment such as pestle and mortar, test-tubes [1]
   - named safety point such as spectacles or gloves [1]
   - naming biuret solution (or substitute) [1]
   - method clearly explained with the need to grind up the piece of food or add water [1], positive result described, turning from blue [1] to a mauve/purple colour [1]

5 Enzymes

> **Overview**
>
> In this chapter, students will investigate factors that affect enzyme activity. They will use their knowledge of enzymes to plan and carry out investigations into how enzymes affect rate of reactions. The results will enhance their knowledge and understanding of how enzyme structure is linked to function.

Practical investigation 5.1 Effect of amylase on starch

Planning the investigation

In this investigation, students will carry out a simple method to observe the effect of amylase on starch. Students will need to carry out the method carefully and observe their results. Students will also suggest an alternative investigation to further test their knowledge of enzymes.

This investigation will focus on the following assessment objectives:

- AO3.1 Demonstrate knowledge of how to safely use techniques, apparatus and materials (including following a sequence of instructions where appropriate)
- AO3.3 Make and record observations, measurements and estimates
- AO3.4 Interpret and evaluate experimental observations and data
- AO3.5 Evaluate methods and suggest possible improvements.

Setting up for the investigation

Prepare 2% starch solution, Benedict's solution and 5% amylase solution in advance and store appropriately. Starch solution can be made by adding 5 g soluble starch to cold water, before stirring into 500 ml of boiling water. Equipment list for each group of 2–4 students: test-tubes × 4, test-tube rack, Benedict's solution, 20 ml of 2% starch solution, 5 ml of 5% amylase solution, Bunsen burner, heat mat, gauze mat, tripod, pipettes × 3, iodine solution, 250 ml glass beaker, safety spectacles, disposable gloves, marker pen.

Use boiled water from a kettle as an alternative to using a Bunsen burner; however, students should reheat the water at the last moment before placing the test-tubes in the water bath. Reuse pipettes by washing them instead of using three pipettes per group to reduce the total number of pipettes required.

Safety considerations

MH **F**

Students should wear safety spectacles and wear gloves (or wash hands) when using an iodine solution. Consult safety data sheets for safe storage and handling of starch solution and amylase. Care must be taken when using Bunsen burners, tripod, gauze mats and kettles. The equipment and glass beaker will remain hot for several minutes and should be allowed to cool before clearing away.

Common errors to be aware of

It is easy to get the test-tubes mixed up if the importance of planning the investigation properly is not emphasised. Encourage students to prepare the results table and label the test-tubes before gathering the rest of the equipment. If students are reusing pipettes, demonstrate their use to them originally, in order to avoid contamination of the solutions.

Supporting your students

The investigation can be simplified by focusing on test-tubes C and D. Omit the relevant stages for this from the method and allow students to observe the direct effect of the starch being broken down into a reducing sugar.

Challenging your students

Ask students to plan a method for the answer to Question 7 in the evaluation section. They can carry out the investigation if you have the time and materials available to do so.

Discussion points and scientific explanation

Test tube	Solutions added	Testing agent used	Colour change
A	2% starch	iodine	no change, remains blue/black
B	2% starch + amylase	iodine	blue/black to brown/red
C	2% starch	Benedict's	no change, remains green/yellow/cloudy
D	2% starch + amylase	Benedict's	green/cloudy – orange/red

Results may vary slightly with differences in concentration, quality and amounts used. The table is a guide to what should be expected. Test-tube A remains the same as a control to show that the change in test-tube B is due to the amylase breaking down the starch. The lack of blue/black colour is due to the absence or low concentration of starch at the end. Test-tube C shows no change for the same reason as amylase was not present to break down the starch. Benedict's solution is the solution used to test for reducing sugar (glucose, in this case) and will show positive, as the amylase has broken the starch down into glucose molecules.

Answers to workbook questions

1. Student completed table
2. Starch is no longer present or is present only at a very low concentration.
3. Answer describes the change from the original colour to the final colour, from yellow/green/cloudy to orange/red
4. Glucose or reducing sugar
5. Orange/red colour after testing with Benedict's solution is a sign of reducing sugar being present.
6. Amylase breaks down starch molecules into glucose molecules. Glucose is a reducing sugar and will turn Benedict's solution from blue to orange/red.
7. For safety reasons or not to burn hands when handling, avoid risk of fire
8. Benedict's will not turn orange/red because the starch will not be broken down. This is because the amylase will be denatured at high temperatures.

Practical investigation 5.2 Effect of temperature on enzyme activity

Planning the investigation

In this investigation, students will carry out an investigation into the effect of temperature on enzyme activity. The method is provided for students but will be followed by an opportunity to practise their graphing skills. The guide is there to help students produce consistent graphs that will help to prepare for examinations.

This investigation will focus on the following assessment objectives:

- AO3.1 Demonstrate knowledge of how to safely use techniques, apparatus and materials (including following a sequence of instructions where appropriate)
- AO3.3 Make and record observations, measurements and estimates
- AO3.4 Interpret and evaluate experimental observations and data.

Setting up for the investigation

1% starch solution and 5% amylase solution will need to be prepared and stored in advance. Equipment list per group of 2–4 students: iodine solution, 20 ml of 1% starch solution, 5 ml of 5% amylase solution, marker pen, pipette × 2, test-tubes × 6, test-tube rack, stopwatch or clock, 250 ml glass beaker × 3, thermometer, safety spectacles, ice cubes, kettle (optional).

Water baths programmed to the three temperatures would be an excellent alternative to making water baths. This will allow for more consistent testing and a direct comparison of student results. Alternatively, a kettle can be used to add a small amount of part-boiled water to room temperature water with the aim of achieving the 35 °C water bath.

Safety considerations

Students to wear safety spectacles and to avoid contact with iodine solution by wearing gloves. If gloves are not available, it is safe to wash hands at the end of the investigation. Consult material safety sheets for the safe storage of the amylase and starch solutions.

Common errors to be aware of

The iodine solution should be fresh or diluted as it may affect the activity of the amylase if it is too strong or not fresh enough. It is best to try the investigation yourself prior to class to ensure that the iodine is suitable for use. If it is not, then your laboratory technician should prepare a fresh bottle.

Supporting your students

These students can draw a labelled diagram of the investigation, showing the six test-tubes in their water baths on the board for easy reference. Diagrams drawn by students that are more able will benefit those students.

Challenging your students

As well as helping with the drawing of the diagrams, these students can experiment with other temperatures to gather more data. For example, testing the enzyme at 50 °C, 70 °C and 85 °C would produce a greater range of results. Students should plot these results as part of their graph.

Discussion points and scientific explanation

Students should identify that the rate of reaction increases as the temperature increases up to 35 °C in this investigation. As the temperature increases, the number of collisions between the enzyme and the starch molecules increases, causing more of the starch molecules to be broken down. Once the starch has been completely broken down, the iodine will no longer be blue/black in colour and will be colourless.

Answers to workbook questions

1. Completed student table
2. Student graph should follow all of the guide points in the student workbook. Look out for axes that are labelled incorrectly or without appropriate units. The guide can also be used as a useful peer/self-assessment exercise for students to check their own progress and identify how the graph could be improved.
3. This should be 35 °C or similar temperature.
4. The starch has been completely broken down (accept almost completely broken down or extremely low concentration of starch left).
5. Student answer should link the two variables of the investigation, using a conditional sentence. 'The rate of reaction increased as the temperature increased' would be a suitable example.
6. Answer should suggest that the enzyme would be denatured at such a high temperature and therefore would not break down the starch. This means that the blue colour would not become colourless.
7. Any sensible suggestion regarding improved control of the control variables

Practical investigation 5.3 Effect of pH on enzyme activity

Planning the investigation

In this investigation, students will carry out an investigation into the effect of pH on enzyme activity, reflect on the quality of the investigation, and link the results to their knowledge of enzyme activity.

5 Enzymes

This investigation will focus on the following assessment objectives:
- AO3.1 Demonstrate knowledge of how to safely use techniques, apparatus and materials (including following a sequence of instructions where appropriate)
- AO3.3 Make and record observations, measurements and estimates
- AO3.4 Interpret and evaluate experimental observations and data
- AO3.5 Evaluate methods and suggest possible improvements.

Setting up for the investigation

Amylase solution, starch solution and a range of buffer solutions can be made/bought in advance but are best prepared fresh for each lesson. Buffer solutions at pH 3, 5, 7 and 9 will give ideal results. The pH can be altered by increasing/decreasing the volume of amylase solution to starch solution. Equipment list for each group of 2–4 students: 10 ml of 1% starch solution, 1 ml of each buffer solution used, 10 ml of the 1% amylase solution, spotting tile, iodine solution, safety spectacles, pipettes, test tubes, test-tube rack, stopwatch.

Safety considerations

Wear safety spectacles at all times. Wear gloves or wash hands after contact with irritants such as iodine solution or amylase solution. Consult safety sheets for safe handling and storage of the enzyme solution, iodine solution and buffer solutions.

Common errors to be aware of

Students may add the solutions in the wrong order, which affects the results. The reaction time might be false if the time is started too slowly or too quickly – teamwork and planning are important for the success of this practical.

Supporting your students

Clear demonstration, supported by a diagram of the spotting tile on the board, will be beneficial to these students. Comparison of just two/three different buffer solutions may simplify the investigation and allow the student to make a more direct comparison between the different pH values.

Challenging your students

Students can calculate the reaction time for this investigation by calculating $1 \div T$. A separate graph can be plotted on graph paper to show these results.

Discussion points and scientific explanation

This will be a similar discussion to the previous investigation, focusing on the action of the enzyme at different pH values. The expected results will be that pH 7 would react the fastest, followed by pH 5, pH 9 and pH 3. pH 3 may not allow the starch to be completely broken down and students should stop testing this once it goes past 10 minutes. Discuss what the optimum temperature is for pH and compare this to the graphs produced by the students.

Answers to workbook questions

1. Completed student table
2. Each drop multiplied by 10 in the final column with the unit denoted as seconds
3. Suitable graph drawn with correctly plotted points
4. Students should identify the optimum pH and observe that the reaction was slower either side of this pH.
5. Answer should describe how the enzyme starts to denature at extreme pH, damaging the active site and preventing further reactions from taking place.
6. Otherwise the reaction would have started before attaining the desired pH
7. Any sensible answer such as repeating the investigation to take an average
8. Accept sensible suggestions such as the use of a colour chart or equipment such as a colorimeter.

Answers to exam-style questions

1 a pH 2.75
   b pH 7.5–8
   c Stomach
   d Mouth, saliva, duodenum or other sensible answer
   e Enzyme would not work, enzyme would denature, active site would be damaged, no reactions would occur, no new products formed, or other similar answer

6 Plant nutrition

Overview

In this chapter, students will investigate factors that affect photosynthesis which will help them better understand the process of photosynthesis. They will calculate the rate of photosynthesis in plants under different conditions and use this data to produce a suitable graph.

Practical investigation 6.1 Testing leaves for the presence of starch

Planning the investigation

In this investigation, students will carry out the basic test for starch. Even though this might be a common investigation, it is important that it is carried out correctly so that it can be used for the planning questions in the rest of this chapter.

This investigation will focus on the following assessment objectives:

- AO3.1 Demonstrate knowledge of how to safely use techniques, apparatus and materials (including following a sequence of instructions where appropriate)
- AO3.3 Make and record observations, measurements and estimates.

Setting up for the investigation

The leaves used should not be too thin or waxy. A soft, thin leaf is ideal and should be exposed to direct sunlight for 4–6 hours in advance. A strong lamp may also be used to shine directly onto the plant before removing leaves.

Buy, or prepare, iodine solution in advance. Any common alcohol solution can be used, such as ethanol, methanol, or even methylated spirits, which may be more readily available in local stores.

Equipment list per group of 2–4 students: Bunsen burner, heat mat, gauze mat, tripod, 250 ml glass beaker, forceps, test-tube tongs, boiling tube, alcohol solution (e.g. ethanol, methanol, methylated spirits), iodine solution, white tile (or Petri dish), leaf, safety spectacles. If Bunsen burners are not used, it is possible to use boiled water from a kettle but the leaf will need to stay in the alcohol for longer until the green colour is removed.

Safety considerations

Wear safety spectacles at all times. Wear gloves or wash hands after contact with iodine solution. Long hair must be tied back and ties removed or tucked in when using the Bunsen burner. The alcohol can explode out of the boiling tube if overheated and so care must be taken to point the end of the tube away from students. The Bunsen burner must be turned off at this stage to avoid ignition of any overheated alcohol. Use a large flask or beaker to collect the wasted alcohol. Consult safety sheets for material and substance preparation and storage.

Common errors to be aware of

Students may not carry out each part of the method for the appropriate time. This is a practical investigation that requires observation of what is happening, rather than relying on strict timings. You could liken this to cooking at home – watching the food to decide when it is cooked, not relying on the timings on the packet. The leaf needs to be plunged into the hot water for long enough to stop all of the chemical reactions inside the leaf and to soften it. The alcohol solution should cover all of the leaf and be left in the alcohol long enough for the leaf to become clear. Some students will remove the leaf after 5 minutes without removing enough of the green colour.

Supporting your students

This method should be broken down into two sections. Demonstrate the preparation of the equipment and placing into the alcohol (step 4) and then direct the students to do the same. While the colour is being removed from the leaf, the rest of the method can be demonstrated and then completed by the students.

Challenging your students

Ask students to make a large drawing of their leaf and calculate the magnification of the leaf. This will encourage them to practise their drawing skills from previous units.

Discussion points and scientific explanation

If the leaf has been in ideal conditions and the method is followed correctly, the iodine solution will turn from a brown/red colour to blue/black in the presence of starch. Ask students to explain how this relates to the equation for photosynthesis and discuss how the plant converts glucose made during photosynthesis to starch for storage.

Answers to workbook questions

1. Any two suggestions from safe storage of alcohol, wash hands after contact with alcohol or iodine, wear disposable gloves, any safe use of Bunsen burner such as carrying safely, safety flame, turn off when not in use, allow to cool after use
2. Student drawing with labels
3. The alcohol was used to dissolve and remove the green chlorophyll from the leaf.
4. The green colour was removed from the leaf to allow the colour change to be viewed when iodine was added.
5. The iodine turned from brown/red to blue/black colour.
6. The darker the colour change, the more starch was present in the leaf.

Practical investigation 6.2 Light as a limiting factor for photosynthesis

Planning the investigation

In this investigation, students will plan their own investigation to observe the effect of light on photosynthesis.
This investigation will focus on the following assessment objectives:
- AO3.1 Demonstrate knowledge of how to safely use techniques, apparatus and materials (including following a sequence of instructions where appropriate)
- AO3.2 Plan experiments and investigations
- AO3.3 Make and record observations, measurements and estimates
- AO3.4 Interpret and evaluate experimental observations and data.

Setting up for the investigation

A suitable plant with thin, non-waxy leaves (ideally, the same type of plant as used in Practical investigation 6.1) should be stored in darkness for at least 48 hours. Remove the plant from darkness, add water and cover parts of the leaves with a strip of aluminium foil and a paper clip. You will need to do this for as many leaves as you will have groups in your class. Do a few extra for your own demonstration and also in case leaves get broken.

Buy, or prepare, iodine solution in advance. Any common alcohol solution can be used, such as ethanol, methanol, or even methylated spirits, which may be more readily available in local stores.

Equipment list per group of 2–4 students: Bunsen burner, heat mat, gauze mat, tripod, 250 ml glass beaker, forceps, test-tube tongs, alcohol solution, (such as ethanol, methanol, methylated spirits), iodine solution, white tile (or Petri dish), boiling tube, safety spectacles, leaf partly covered with aluminium foil, paper clip
If Bunsen burners are not used, it is possible to use boiling water from a kettle but the leaf will need to stay in the alcohol for longer until the green colour is removed.

Safety considerations

Wear safety spectacles at all times. Wear gloves or wash hands after contact with iodine solution. Long hair must be tied back and ties removed or tucked in when using the Bunsen burner. The alcohol can explode out of the boiling tube if overheated and so care must be taken to point the end of the tube away from students. The Bunsen burner must be turned off at this stage to avoid ignition of any overheated alcohol. Use a large flask or beaker to collect the wasted alcohol. Consult safety sheets for material and substance preparation and storage.

Common errors to be aware of

As in Practical investigation 6.1, students may be too hasty with the timings of the investigation. Reaffirm the need to observe the changing colour of the leaf to avoid student error and overly green leaves. The plant must receive sufficient sunlight to begin reproducing starch and so may benefit from being placed under direct light overnight.

Supporting your students

Plan the investigation as a teacher-led small group while others work independently. Draw the method out using questioning and only add to the method once the correct step has been suggested.

Challenging your students

Students can plan a similar investigation and make predictions if carbon dioxide was the limiting factor removed from the investigation.

Discussion points and scientific explanation

Discuss the results with students and explain that the starch was completely removed from the plant during the 48 hours of darkness. The aluminium foil prevented light entering the covered parts of the leaf when reintroduced to the light and so those areas did not photosynthesise. As photosynthesis did not occur, glucose was not produced and therefore no glucose was converted to starch. The covered areas remained yellow/pale in colour while the exposed parts turned blue/black with iodine solution.

Answers to workbook questions

1. Benedict's solution, conical flask and thermometer are not needed.
2. Student method
3. Outline of safety precautions
4. Student drawing
5. It remained yellow/orange in colour, no colour change.
6. This change happened because there was no starch present as glucose could not be converted to starch as there was no glucose produced in those areas.
7. To remove all starch from the plant and observe the effect of light on the leaf when reintroduced to the light

Practical investigation 6.3 Effect of light intensity on oxygen production in Canadian pondweed

Planning the investigation

In this investigation, students will investigate the effect of light intensity on the rate of photosynthesis by observing the amount of oxygen released by the plant. This investigation will focus on the following assessment objectives:

- AO3.1 Demonstrate knowledge of how to safely use techniques, apparatus and materials (including following a sequence of instructions where appropriate)
- AO3.3 Make and record observations, measurements and estimates
- AO3.4 Interpret and evaluate experimental observations and data.

Setting up for the investigation

Canadian pond weed (*Elodea canadensis*) is ideal for this but most underwater plants will work. The plant can be stored under water for several days in advance and the more leaves on the pondweed, the better the results will be.

Equipment list per group of 2–4 students: pieces of water plant, boss clamp and stand, light source/lamp, stopwatch, boiling tube, metre ruler, paper clip. Adding a paper clip (or two) will anchor the water plant for this investigation and allow space for the bubbles to be observed at the top of the boiling tube.

Safety considerations

Take care when using the lamp as this will become very hot after a few minutes. Encourage students to turn the lamp off as soon as the investigation has concluded.

Common errors to be aware of

The ends of the stem of the water plant can close up. If this happens, the rate of bubbles will slow down. The investigation can be broken up into stages to allow the plant to be removed and have the ends re-cut. The concentration of the carbon dioxide dissolved in the water can be increased by adding a few grams of sodium hydrogencarbonate to the water.

Supporting your students

Reduce the number of distances measured for the investigation to make it simple. Assign roles to students within small groups to ensure that the number of bubbles is not missed or counted incorrectly. Students may need some help with the calculation of the mean number of bubbles – either in small groups or individually.

Challenging your students

Ask students to research and suggest a method for collecting the gas produced and testing it to determine its identity.

Discussion points and scientific explanation

Discuss how light was the limiting factor in this investigation. Identify the point on the graph at which the rate of bubbles produced no longer increased and light was no longer the limiting factor. Extend the discussion to suggest using a gas syringe to collect the gas and measure the rate more accurately. Ask students to describe how to test the gas to identify it as oxygen – oxygen will relight a glowing splint.

Answers to workbook questions

1. Completed table
2. Answer should describe the addition of the three values for minutes 1, 2 and 3 before dividing this total by the number of measurements.
3. Suitable line graph drawn following the guidelines, with a smooth curve drawn through the points plotted
4. Student answer should describe the first part of the curve as increasing at a steady rate before becoming a straight line/levelling off/straightening out at a certain distance between the plant and the light source.
5. Light was no longer a limiting factor.
6. To improve reliability, to spot anomalies in the data or method

Answers to exam-style question

1.
   a. Answer shows number of whole squares and part squares counted [1], correct answer ± 2 cm^2 [1]
   b. Diagram is the same size or larger than the original [1], the leaf is drawn with smooth lines [1], no shading [1], in the correct shape and proportion [1], observable features of the leaf are correctly labelled [1]
   c. Magnification = image size/actual size [1], answer correct based on student drawing and greater than 1 [1]
   d. The inside, green area of the leaf [1]
   e. Photosynthesis has happened here [1] as there is chlorophyll present, thus producing glucose, which is stored as starch [1]. Starch turns blue/black in the presence of iodine [1].

7 Animal nutrition

> **Overview**
>
> In this chapter, students will investigate the energy contained in different types of food and how physical and chemical digestion take place in the human body.

Practical investigation 7.1 Measuring the energy content of foodstuffs, part I

Planning the investigation

Students will set up an investigation to measure the energy content of different foods. The investigation has a basic method in the book with many areas for poor reliability; students will either notice these and amend accordingly or will carry out the investigation without managing the control variables in the investigation. The evaluation stage will guide the planning of Practical investigation 7.2 that follows.

This investigation will focus on the following assessment objectives:
- AO3.1 Demonstrate knowledge of how to safely use techniques, apparatus and materials (including following a sequence of instructions where appropriate)
- AO3.2 Plan experiments and investigations
- AO3.3 Make and record observations, measurements and estimates
- AO3.4 Interpret and evaluate experimental observations and data
- AO3.5 Evaluate methods and suggest possible improvements.

Setting up for the investigation

Purchase a wide range of food samples in advance that are suitable for burning, such as biscuits, pasta, crisps, chips, bread, chocolate, nuts and candy. Where possible, you could ask students to bring samples from home to maximise their engagement with the investigation. Equipment list per group of 2–3 students: Bunsen burner, boss clamp and stand, boiling tubes × 4, thermometer, heat mat, safety spectacles, mounted needle, food samples.

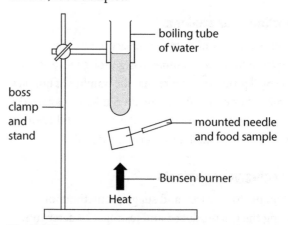

Figure 7.1

See Figure 7.1. A tripod and crucible can be used as an alternative to the boss clamp and stand; the water could be heated in a glass beaker. As long as the food is between a stable source of water and the Bunsen flame, the results will be sufficient. Reuse boiling tubes instead of allocating four per group to reduce the number of boiling tubes required.

Safety considerations

Wear safety spectacles at all times, secure all loose clothing, and tie long hair back. The mounted needle and boiling

Cambridge IGCSE Biology

tubes get very hot and extra care must be taken to avoid burning hands. The water will erupt from the boiling tube if boiled – tell students to cease burning the food if this happens. If necessary, the food can be retested rather than risk injury. The boiling tube should be pointed at an angle using the clamp and aimed away from students.

Common errors to be aware of

Students may place the flame too close to the boiling tube and thus the flame rather than the heat energy released from the food heats the water. Avoid this by encouraging students to trial the investigation before recording data. Ask them if they are happy with the distance and amend the distance between boiling tube and Bunsen flame if necessary.

Supporting your students

This is another investigation where a struggling student will benefit from collecting less data and carrying out the method in stages alongside the teacher. Allow most students to go ahead with the method while you demonstrate systematically how to set up the investigation.

Challenging your students

Direct students to find out how to calculate the energy change using $Q = mc\Delta T$. Students will need to look up the specific heat capacity of water, which is 4.2 J/Kg °C.

Discussion points and scientific explanation

Discuss with students how the food that caused the greatest change in the temperature of the water released the most heat energy. Discuss which food types had the most and least energy. The focus of this investigation is to highlight and improve areas of reliability. Students should answer the questions in the evaluation stage and discuss the options with you. Either you can plan a new method for Practical investigation 7.2 using this information or it can be done individually or in groups.

Answers to workbook questions

1. Completed table, with units °C
2. Student answer should match the food sample in their table with the greatest change in water.
3. The temperature change of the water was greater than other food samples. Student should confirm this by quoting the actual temperature change of that food.
4. Test each food more than once and calculate the mean temperature change.
5. Measure the same volume of water for each sample of food tested with a measuring cylinder.
6. Use the same size/shape and/or mass for each of the food samples used.
7. Stir the water to get a consistent temperature and avoid touching the sides of the boiling tube with the thermometer.

Practical investigation 7.2 Measuring the energy content of foodstuffs, part II

Planning the investigation

In this investigation, students will use their evaluation suggestions from Practical investigation 7.1 to plan a more reliable investigation into the energy content of food. This investigation will focus on the following assessment objectives:

- AO3.1 Demonstrate knowledge of how to safely use techniques, apparatus and materials (including following a sequence of instructions where appropriate)
- AO3.2 Plan experiments and investigations
- AO3.3 Make and record observations, measurements and estimates
- AO3.4 Interpret and evaluate experimental observations and data
- AO3.5 Evaluate methods and suggest possible improvements.

Setting up for the investigation

This is the same as for Practical investigation 7.1 but with the addition of extra equipment for each group to improve reliability. In addition to the equipment used before, each group may/will also require a 100 ml measuring cylinder, balance or weighing scales suitable for small masses, and possibly a knife or scissors for cutting the food samples to the desired size/mass. Extra boiling tubes will be required or students should wash/cool the boiling tubes between tests.

Safety considerations

Wear safety spectacles, tie hair back, secure loose clothing, turn off the Bunsen burner when not in use, and take extra care when handling the hot mounted needle and boiling tubes.

Common errors to be aware of

There is a lot of data to collect and errors can be incurred if students rush trying to complete within one lesson. Allow students to complete the investigation over two lessons if necessary.

Supporting your students

Plan the new investigation together as a teacher-led group and offer support to students for the calculations involved. Carry out calculations as a class/group once the investigation has been completed.

Challenging your students

Students can lead the class, or small group, discussion of the calculations. Select some students to support the less able students with calculations.

Discussion points and scientific explanation

Discussion should focus on the improved reliability and opportunities for improvements in accuracy. Some discussion and comparison of the energy content of foods may be carried out to reinforce knowledge of food types with high energy content.

Answers to workbook questions

1. Student equipment lists
2. Student methods
3. Student safety precautions
4. Student answer should match food with greatest energy change.
5. Student describes the temperature change and/or the heat energy transferred in the investigation, giving figures from their calculations.
6. Students' own answer
7. Students' own answer, to support their answer to Question 6
8. Any sensible answer such as using a digital thermometer with more decimal places for increased accuracy

Practical investigation 7.3
Mouthwash versus acids

Planning the investigation

In this investigation, students will compare the difference between how acids and mouthwash affect our teeth. An egg is used to represent the teeth and the drawing skills of students can be improved as they record their observations.

This investigation will focus on the following assessment objectives:

- AO3.1 Demonstrate knowledge of how to safely use techniques, apparatus and materials (including following a sequence of instructions where appropriate)
- AO3.3 Make and record observations, measurements and estimates.

Setting up for the investigation

The eggs need to be hard-boiled in advance. The number of eggs needed depends on whether you want to provide pairs or small groups with eggs, or just provide two eggs for the class to observe together. Once placed into their respective solutions, the eggs will need to be left for at least 24 hours. Allow two lessons to carry out this investigation. Alternatively, pre-prepared eggs can be placed in solution and shown to the students without them having to prepare the eggs in the jars.

Equipment list per group: hard-boiled eggs × 2, jars with screw-on lids × 2, 100 ml vinegar (or enough to submerge an egg), and 100 ml fluoride mouthwash (or enough to submerge an egg), marker pen, disposable gloves, safety spectacles.

Safety considerations

MH

Safe disposal of the solution once the egg is removed. Wear gloves when removing the egg from the solution. Wear safety spectacles.

Challenging your students

Ask students to take photographs of their eggs and use these to produce a poster to promote good care and hygiene of teeth.

Discussion points and scientific explanation

Discuss how fluoride mouthwash works against the acids that come into contact with our teeth to keep our teeth healthy and strong.

Answers to workbook questions

1. Student's drawing
2. Answer suitably describes the differences seen in the eggs. The egg in the mouthwash will remain smooth and look stronger than the cracked eggshell from the egg that was placed in the acid.
3. Acids dissolve the enamel of the teeth, exposing the dentine. The dentine may also dissolve and severe toothache and pain will be experienced.
4. Fluoride helps to make your teeth resistant to decay and protects the enamel from being dissolved.
5. To ensure a reliable test/comparison

Answers to exam-style questions

1. a Degrees Celsius or °C [1]
   b 32.5, 72, 43.5 [1]
   c 8, 48, 19 [3]
   d Any relevant safety suggestion such as wear safety spectacles, tie hair back, secure loose clothing, do not handle hot Bunsen burner or other named equipment [1]
   e Turned from blue [1] to orange/red [1]
   f Carbohydrates, accept fats [1]
2. a i Enamel [1]
     ii Cementum [1]
     iii Root [1]
   b Sugar in drink is absorbed by plaque [1], bacteria convert plaque to acid [1], which dissolves the enamel layer of the tooth [1], leading to tooth decay and infection [1].
   c Brushing teeth every day [1], using fluoride mouthwash (or other named product) [1], reduce consumption of sugar [1], regular visits to the dentist [1], and any other relevant suggestions may be accepted.

8 Transport in plants

> **Overview**
>
> In this chapter, students will investigate how water moves in, out and through a plant. They will use their results to calculate the rate at which water enters and exits a plant.

Practical investigation 8.1
Transport of water through plants via xylem

Planning the investigation

In this investigation, students will carry out a simple method to observe how water travels up through the stem of the plant. Students will then use this to practise their drawing skills and repeat the calculations required for magnification from previous chapters.

This investigation will focus on the following assessment objectives:

- AO3.1 Demonstrate knowledge of how to safely use techniques, apparatus and materials (including following a sequence of instructions where appropriate)
- AO3.3 Make and record observations, measurements and estimates
- AO3.5 Evaluate methods and suggest possible improvements.

Setting up for the investigation

Buy as much celery as required for your class. You can give each group their own stem or, if your class is larger, you can provide one stem per five or six groups. These groups can then share the celery between them when slicing into cross-sections.

Equipment per group of 1–4: celery stem with leaves attached, food colouring/dye, safety spectacles, scalpel/knife, 250 ml glass beaker, stopwatch/clock, pipette.

1% methylene blue can also be used but it is usually easier to use any food colouring or dye; red or blue tends to give the clearest results. Buy the celery as fresh as possible and store in the refrigerator overnight if necessary.

Safety considerations

Students to wear safety spectacles as standard practice. Take care when handling and using the knife or scalpel.

Common errors to be aware of

It is possible that the stem is not left in the coloured water for long enough. The simple way to ensure that you get the desired results is to observe the leaves. When the leaves are showing the colour of the dye, you are sure to have coloured xylem vessels.

Supporting your students

This is a straightforward investigation but some students might need reminding or scaffolding for the drawing and magnification sections. A question and answer session before starting the investigation should be sufficient to remind students of expectations.

Challenging your students

Ask students to make a labelled drawing of the leaves. They can calculate the magnification of these also.

Discussion points and scientific explanation

The coloured vessels show where the water travelled through the xylem vessels. Water moves from the roots to the leaves via the stem. Remind students that water is absorbed through the roots and is needed for photosynthesis.

Answers to workbook questions

1. Student drawing
2. Student magnification
3. Xylem vessel
4. Xylem vessels are lined with lignin, which is rigid and strong.
5. Minerals, mineral salts
6. Tests more than one type of plant, shows that it is not just celery that the movement of water occurs in
7. To observe the passage of water, clear water would not be visible/observable otherwise

Practical investigation 8.2 Testing the product of transpiration

Planning the investigation

In this investigation, students will observe and confirm the identity of the product released by transpiration in plants. Students will plan an investigation to compare the different rates between two different plants in preparation for Practical investigation 8.3.

This investigation will focus on the following assessment objectives:

- AO3.1 Demonstrate knowledge of how to safely use techniques, apparatus and materials (including following a sequence of instructions where appropriate)
- AO3.2 Plan experiments and investigations
- AO3.3 Make and record observations, measurements and estimates
- AO3.5 Evaluate methods and suggest possible improvements.

Setting up for the investigation

Advance purchase of any leafy, potted plant found in your local area will be required. The geranium (*Pelargonium*) is a suitable example and can be used as a guide for the sort of plant required. Preparation of anhydrous copper sulfate is needed with the appropriate storage as recommended in your safety sheet guidelines. If anhydrous copper sulfate is not available, cobalt chloride paper can be used or made. Pupils can set up and observe the plant across multiple lessons if required. Advance set-up by the teacher or the technician will allow students to focus solely on the results; do this by carrying out steps 1 and 2 in advance of the lesson. Inform students that steps 1–2 were prepared in advance for them. Equipment: potted plant, clear polythene bag, string/cable tie, anhydrous copper sulfate, spatula, pipette, safety spectacles.

Safety considerations

Wear safety spectacles and follow safety sheet guidelines for use of anhydrous copper sulfate. Wash hands if in contact with this substance.

Common errors to be aware of

Take care not to spill the water collected in the bag or to allow soil to fall back into the bag if tilting the plant. The tie (which can be string, cable tie, wire or other suitable material) must secure the bag fully to the stem of the plant and not allow the water to escape.

Challenging your students

Students can calculate the volume of water lost due to transpiration by using a measuring cylinder or calculating the mass and converting to volume.

Discussion points and scientific explanation

Recap the definition of transpiration and discuss how anhydrous copper sulfate is a true test for the presence of water. The structure of the leaf – waxy layer – relates to minimising the loss of water as a recap of prior knowledge.

> **Answers to workbook questions**
>
> 1 Student observations
> 2 Student observations
> 3 It turns blue, which only happens to anhydrous copper sufhate in the presence of water.
> 4 Loss of water vapour from plant leaves by evaporation of water at the surfaces of the mesophyll cells followed by diffusion of water vapour through the stomata
> 5 Investigation should outline a similar method to Practical investigation 8.2 but by using two different types of plant. Answer may identify differences between the leaves as a factor, control variables should be identified such as plant size, surrounding conditions, and size and type of polythene bag.

Practical investigation 8.3
How environmental factors affect the rate of transpiration

Planning the investigation

In this investigation, students will use a potometer to measure the rate of water uptake by a plant and link this to the rate of transpiration for that plant. They will test this under different conditions, draw a graph, and draw conclusions from their results.

This investigation will focus on the following assessment objectives:
- AO3.1 Demonstrate knowledge of how to safely use techniques, apparatus and materials (including following a sequence of instructions where appropriate)
- AO3.3 Make and record observations, measurements and estimates
- AO3.5 Evaluate methods and suggest possible improvements.

Setting up for the investigation

If a potometer is available, the only advance set-up will be the purchase of a suitable leafy plant similar to those used in Practical investigation 8.2. A potometer can be made by fixing one end of a glass tube to the cut end of the plant stem/shoots. Air gaps must be sealed by a substance such as petroleum jelly and the movement of water can be measured by using the scale, or a ruler. A reservoir of water can also be used by taking the volume before and after to find the final volume of water that was taken in by the plant. There are many different potometers that can be bought, or made; an internet search engine will provide many diagrams for you to be guided by.

Equipment per group of 2–4 students: potometer, leafy plant, electric fan (this can be shared between groups), ruler, petroleum jelly (or similar substance).

Safety considerations

Take care when fixing the tube to the plant. Wash hands after handling the plant.

Common errors to be aware of

Careful set-up eliminates most errors here, avoiding the air gaps previously mentioned. The rate of transpiration can take longer to slow down than to speed up. It may be better to begin with the plant in the dark area rather than in the sunlight.

Challenging your students

Students can investigate other areas or conditions such as change in humidity or temperature. They can either carry this out or simply plan a suitable method.

Discussion points and scientific explanation

Students discuss how the rate changed and under which conditions. Lead discussion towards which conditions cause the stomata to open and increase the loss of water – such as an increase in light intensity or temperature. The speed of the fan will also affect the rate, as more water is lost as the rate of evaporation increases. Stomata will open under increased light intensity as the plant will require more carbon dioxide to enter for photosynthesis.

Answers to workbook questions

1. Student table
2. Student graph shows time (minutes) on the x-axis and water uptake (ml) on the y-axis. Points plotted correctly and a straight line drawn with a pencil and ruler. The three lines should be in different colours or clearly marked with a key to denote the different conditions.
3. Student answer matches the data from their graph/table – likely to be the fan.
4. Student answer makes relevant connection between the result and the reason behind it. The fan caused a greater uptake of water as the windy conditions led to an increase in evaporation. This leads to an increase in water taken up by the roots.
5. Transpiration causes the water to be 'pulled' up through the xylem so an increase in transpiration leads to an increase in the amount of water being taken in at the roots.
6. The conditions that the plant was placed in
7. The amount of water taken in by the plant (per minute)
8. The same plant used each time, same amount of water in the reservoir/tubing, same conditions other than the one being changed each time, any other sensible answer

Answers to exam-style questions

1. Student answer describes the molecule of water being taken into the root [1] by root hair cells [1] before moving through the xylem [1] to the leaf where it will leave the leaf through the stomata [1] by evaporation [1].
2.
   a. Root hair (cell) [1], large surface area to absorb as much water as possible, thin cell wall [1] for shorter diffusion pathway [1] for water molecules
   b. Osmosis is the net movement of water molecules [1] from an area of higher water concentration to an area of lower water concentration through a semi-permeable membrane (or down the concentration gradient) [1], in this case water moves from an area of higher water concentration in the soil to an area of lower water concentration inside the cell [1].
   c. Leaf, root, flower

9 Transport in animals

Overview

In this chapter, students will observe how the heart and lungs are structured, and how they contribute to the well-being of the body by supplying organs with the substances that they require. Students will link their knowledge of these organs to how they respond under different conditions.

Practical investigation 9.1
Dissecting a heart

Planning the investigation

In this investigation, students will dissect the heart of a mammal and use this to identify the different parts of the heart.
This investigation will focus on the following assessment objectives:
- AO3.1 Demonstrate knowledge of how to safely use techniques, apparatus and materials (including following a sequence of instructions where appropriate)
- AO3.3 Make and record observations, measurements and estimates
- AO3.5 Evaluate methods and suggest possible improvements.

Setting up for the investigation

You need to purchase as many hearts as required for your class. Ideally, each student will have a heart each but it may be that you allow one per pair. Purchase your hearts from your local science supplier or your local butchers. Your local supermarket will also have a range of hearts from which to choose. The best hearts are those of a sheep or a lamb but any animal of similar size would be suitable. Avoid the use of a pig's heart, as this may be culturally insensitive to some students. The heart should be bought as fresh as possible but can be preserved in certain substances, such as formaldehyde, for up to a year. Prior to the investigation, consult with your local science supplier for the availability of these in your area. Your local butcher, and some butchers in supermarkets, will be able to provide hearts to order with as many of the vessels intact if requested in advance. Explain to students about missing vessels if this is not possible and direct them to suggest what is missing as part of their understanding.

A dissection kit should be used if available but a sharp scalpel and surgical scissors will suffice. Using a mounted needle can help to control and probe the heart.
Equipment list per 1–2 students: mammal heart, scalpel, mounted needle, surgical scissors, white tile or dissection tray, latex gloves, paper towels, apron(s), safety spectacles, pipette or syringe, forceps, water.
If using a dissection tray or similar, the equipment can be set up per tray. Otherwise, allow students to select the items that they require.

Safety considerations

Wear safety spectacles, gloves and an apron at all times. Wash hands after the investigation. Do not eat or drink in the laboratory – this should be the case for all investigations, in this case to avoid ingestion of airborne pathogens. Take care when handling the scalpel and scissors. It is important that you demonstrate how to make gentle slices with the scalpel, allowing the blade to do the work rather than trying to force the knife through the heart. Clean all work surfaces with disinfectant after the class has finished and dispose of all hearts following your local waste disposal guidelines.

Common errors to be aware of

The most common problem here is a lack of confidence. Guide students step-by-step to make decisive incisions and not to get frustrated or flustered if the heart does not slice open easily. The 1 cm removal method is easier and students will find this satisfying as they get instant results. Some students will risk injury by trying to 'hack' through the heart. Encourage students to use the scalpel to slice slowly through the heart gradually. Students should also use the surgical scissors to cut through the heart.

Supporting your students

As above, using the surgical scissors to cut through the heart after being shown by the teacher is easier for some students than delicately opening the heart with the scalpel, which is more difficult. Offering a choice of methods will boost confidence as students can choose what they think is the best way to dissect the heart.

Challenging your students

Direct students to label the parts of the heart with small dissection pins. Use sticky labels to make 'flags' that can be pinned to the heart to identify the part of the heart being observed. If students have access to hypodermic needles, direct them to inject a blue dye into the coronary artery at a shallow angle to follow the path of the artery across the heart wall.

Discussion points and scientific explanation

Most of the discussion will take place during the demonstration. Ask students to compare the strength and thickness of the aorta with the pulmonary vein and discuss the difference in these structures. You may also demonstrate/simulate the passage of blood through the heart using water and pipette/syringe to encourage identification of the different chambers of the heart.

Answers to workbook questions

1 Student labelled drawing
2   a To prevent bacteria coming into contact with skin or clothes
    b To destroy or remove any bacteria that may be on the hands or skin
    c To destroy any bacteria that remain on the surface and prevent this bacteria coming into contact with anyone who touches the work surface
    d To prevent airborne pathogens from being ingested

Practical investigation 9.2 Effect of exercise on heart rate, part 1

Planning the investigation

In this investigation, students will carry out an investigation to measure their heart rate after different types of exercise. Students will evaluate their investigation and make suggestions on how to improve the accuracy of the results.

This investigation will focus on the following assessment objectives:

- AO3.1 Demonstrate knowledge of how to safely use techniques, apparatus and materials (including following a sequence of instructions where appropriate)
- AO3.3 Make and record observations, measurements and estimates
- AO3.4 Interpret and evaluate experimental observations and data
- AO3.5 Evaluate methods and suggest possible improvements.

Setting up for the investigation

Identify a suitable area for exercise. This could be an outdoor area depending on the weather, or a large room or hall if you have one available. Students can do on-the-spot exercises such as star-jumps, squats, walking and jogging if space is limited.

Equipment required per group of 2–4 students: space, stopwatch.

Safety considerations

Awareness of the area should be assessed and communicated to the students. Explain to students that they should stop exercising if they feel unwell or dizzy.

Common errors to be aware of

Students not allowing their pulse to return to their resting rate. Direct students to rotate the activity so that one person is always resting while another carries out the activity.

Supporting your students

Assign a more able student to each group to take on the role of manager or group leader. Direct this student to organise students into their activities. This will allow the struggling students to focus on what they are doing rather than the order of who does what.

Challenging your students

Encourage the more able students to support struggling students as described above. Encourage students to think about how they break the method down so that it is clear to their peers.

Discussion points and scientific explanation

Compare the results for different students and discuss the need for increased blood supply during and after exercise. The body requires more oxygen and glucose to be transported in order to do more respiration to release the energy required for the increase in activity. Ask students to suggest reasons why some students had a much greater increase in heart rate during and after exercise. Link this to elite athletes who can often have a lower resting heart rate and therefore their heart does not have to work as hard to supply the needs of the body.

Answers to workbook questions

1. Students' populated table
2. Student describes, using data from their table, the increase in heart rate from resting to light exercise to heavy exercise.
3. The body requires more energy from respiration and so more oxygen and glucose is needed in the cells in order to meet this demand.
4. Oxygen and glucose are transported in the blood so the rate at which blood is pumped must increase to meet the demand.
5. Any sensible suggestion of technology, such as pulse monitor, pulse oximeter, electrocardiography, etc.

Practical investigation 9.3 Effect of exercise on heart rate, part II

Planning the investigation

In this investigation, students will carry out an investigation to measure their heart rate after several minutes of exercise. Students will use their results to plot their results as a line graph.

This investigation will focus on the following assessment objectives:

- AO3.1 Demonstrate knowledge of how to safely use techniques, apparatus and materials (including following a sequence of instructions where appropriate)
- AO3.2 Plan experiments and investigations
- AO3.3 Make and record observations, measurements and estimates
- AO3.5 Evaluate methods and suggest possible improvements.

Setting up for the investigation

Identify a suitable area for exercise. This could be an outdoor area depending on the weather, or a large room or hall if you have one available. Students can do on-the-spot exercises such as star-jumps, squats, walking and jogging if space is limited.

Equipment required per group of 2–4 students: space, stopwatch.

Safety considerations

Awareness of the area should be assessed and communicated to the students. Explain to students that they should stop exercising if they feel unwell or dizzy.

Supporting your students

Remind students of the steps for planning a method, making a table and plotting a graph. Display the variables of the investigation and discuss where they will be placed on the table and the graph.

Challenging your students

Ask students to calculate the percentage increase in their heart rate and discuss the reasons behind the different changes for each person. There is an opportunity here for students to practise their microscope skills by looking at pre-prepared slides that contain different blood cells. Direct students to examine these and make a biological drawing; using the similar method shown in practical investigation 2.3.

Discussion points and scientific explanation

Compare the results for different students and discuss the need for increased oxygen/glucose during and after exercise. The body requires more oxygen/glucose in order to do more respiration to release the energy required for the increase in activity. Ask students to suggest reasons why some students had a much greater increase in heart rate during and after exercise.

Answers to workbook questions

1. Awareness of safe area and space, student to monitor well-being of those exercising.
2. Student suggests a sensible method to record the heart rate by measuring the number of beats per minute at different time intervals.
3. Student table has the heart rates (beats per minute) recorded at the stated time intervals.
4. Student graph plots the data for time (minutes) on the x-axis and heart rate (beats per minute) on the y-axis.
5. Heart rate increases as time exercising increases.
6. The body requires more oxygen/glucose for respiration so the heart rate increases to increase the amount of oxygen/glucose being supplied to cells.

Answers to exam-style question

1.
   a To wash away bacteria [1] that may cause infection of the wound [1]
   b This forms a 'scab' over the cut [1], preventing further blood loss [1]
   c Fibrin [1]
   d Slows the red blood cell down [1] while travelling through the capillaries to allow more substances to move in and out of the cell [1], such as oxygen, carbon dioxide or any other correctly named substance [1]
   e Any three from: amino acids, lipids, glycerol, sugars, glucose, hormones, heat energy, proteins [3]

10 Pathogens and immunity

Overview

In this chapter, students will investigate the bacteria which are present in their surroundings and link this to how the human body protects against pathogenic organisms entering the body. Knowledge of these procedures is important in protecting ourselves from pathogenic organisms that are around us.

Practical investigation 10.1 Culturing bacteria

Planning the investigation

In this investigation, students will cultivate bacteria using a nutrient broth to observe the rapid reproduction and potential danger of pathogens. The focus of this unit is safety as microbiological techniques are potentially dangerous if the appropriate safety guidelines are not checked and used.

The safety precautions for this unit are too lengthy and varied to be summarised in this book. It is recommended that you consult your safety data sheets, local regulations and a microbiology safety guide for the materials and substances that you use.

This investigation will focus on the following assessment objectives:

- AO3.1 Demonstrate knowledge of how to safely use techniques, apparatus and materials (including following a sequence of instructions where appropriate)
- AO3.3 Make and record observations, measurements and estimates.

Setting up for the investigation

Advance purchase and storage of the following must be made from your local scientific supplier.

- Sterile agar plates. (Your local doctor or hospital may also be willing to donate a few sterile plates in some cases.)
- Microbe sample – the selection of this is dictated by your available supplier. Bacteria types that have been successful include *Staphylococcus epidermidis*, *Escherichia coli*, *Micrococcus luteus*, and *Staphylococcus albus*. There are many others available but you must consult your supplier and microbiological guidelines before ordering and using. Observe the shelf-life and safety guidelines supplied by the supplier.
- Sticky tape must be clear otherwise results will be obscured. Use Biohazard tape (or similar) if you have this available and place this over the edges of the petri dish only. This will ensure that the results are easily visible.

The method suggests that students clean their benches with disinfectant; however, this can be done by the technician before the lesson starts if possible.

Equipment list per group of 2–4 students: disinfectant, safety spectacles, sterile agar plate, marker pen, Bunsen burner and heat mat, liquid microbe sample, inoculating loop, sticky tape, access to an incubator at 25 °C. If you do not have access to an incubator, leave the dishes at room temperature in a safe place such as a lockable prep room.

Safety considerations
MH **HH**

You must consult your safety data sheets and guidelines as well as familiarise yourself with basic microbiology precautions relevant to the materials being used. Prior to carrying out the investigation, consult with

the laboratory technician or teacher with relevant microbiological experience. Adhere to the following safety rules at all times:
- Wear safety spectacles.
- Do not open the Petri dish once it has been sealed.
- Do not eat or drink in the laboratory.
- Do not ingest or inhale any of the substances at any time.
- All open cuts or wounds must be covered.
- The Petri dishes must be autoclaved at a minimum of 120 °C before safe disposal.

Common errors to be aware of

Students will sometimes fail to sterilise the inoculating loop for long enough but this will not have a noticeable effect on results. Students will either spread the loop on the agar too softly, or too hard. This leads to the bacteria not being spread around or the agar being torn up. Practise this technique in advance and provide a clear demonstration for the students.

Supporting your students

Stagger the start times for some students so that you can work closely with them for the techniques. Direct students to practise the spreading technique using a pencil and paper. They can draw the circle of a Petri dish and show you the right amount of pressure by using the pencil on the paper. Too hard and the pencil line will be too heavy; too light and the pencil will not be visible.

Challenging your students

Ask students to prepare a 'Guide to microbiology' with the relevant safety steps explained.

Discussion points and scientific explanation

Divide the discussion into two main parts – the safety and the results. Discuss safety at the end of the first lesson and link to how airborne pathogens are a danger. Demonstrate this by placing cotton wool soaked in water inside a sealed canister with two frozen peas on top of the wool. Leave these for a week at room temperature and observe the colonies of bacteria and fungi that have grown on the peas. Discuss how the pathogens must have been airborne to reach the peas. Obviously, this will need to be prepared in advance of this discussion. Alternatively, prepare this as part of the lesson and revisit in a week's time. Discuss the use of high temperatures from the Bunsen flame and the use of disinfectant to destroy bacteria and create a sterile environment for the investigation.

When the students observe the results, discuss the reproduction rates of the bacteria and how they can multiply very quickly. Depending on the number of bacteria, it can take a long time to count all of the colonies. Demonstrate how students can count a section of the dish and scale this up to estimate the full bacterial colony count on the dish. Discuss how the incubator and nutrient broth contribute towards the ideal conditions for bacterial growth.

Answers to workbook questions

1. Student drawing
2. Description of counting a smaller area and multiplying to estimate the total number of colonies
3. Student answer matches their Petri dish
4. Incubation at 25 °C, use of nutrient broth, sealed container, away from agents that destroy or inhibit bacterial growth, or any other sensible suggestion
5. Bacteria would not grow.
6. To provide a sterile environment for the investigation
7. Bacteria would multiply too quickly, dangerous.
8. The airborne bacteria that would be released would be very dangerous for anyone who ingested or inhaled them. Opening the dish may also expose bacteria to different conditions and affect the results seen.

Practical investigation 10.2
Bacteria around you

Planning the investigation

In this investigation, students will use the skills from culturing bacteria in Practical investigation 10.1 to investigate the bacteria from their natural surroundings. As much of the content is similar to Practical investigation 10.1, the following guidelines may refer you back to that investigation.

The safety precautions for this investigation should be drawn from your safety data sheets, local regulations and a microbiology safety guide for the materials and substances that you use.

This investigation will focus on the following assessment objectives:
- AO3.1 Demonstrate knowledge of how to safely use techniques, apparatus and materials (including following a sequence of instructions where appropriate)
- AO3.3 Make and record observations, measurements and estimates.

Setting up for the investigation

Make an advance purchase and storage of sterile agar plates from your local scientific supplier.

The method suggests that students clean their benches with disinfectant; however, this can be done by the technician before the lesson starts if possible.

Equipment list per group of 2–4 students: disinfectant, safety spectacles, sterile agar plate, marker pen, Bunsen burner and heat mat, inoculating loop, sticky tape, access to an incubator at 25 °C. If you do not have access to an incubator, leave the dishes at room temperature in a safe place such as a lockable prep room.

Safety considerations

MH

You must consult your safety data sheets and guidelines as well as familiarising yourself with basic microbiology precautions relevant to the materials being used.

Prior to carrying out the investigation, consult with the laboratory technician or teacher with relevant microbiological experience. Adhere to the following safety rules at all times:
- Wear safety spectacles.
- Do not open the Petri dish once it has been sealed.
- Do not eat or drink in the laboratory.
- Do not ingest or inhale any of the substances at any time.
- All open cuts or wounds must be covered.
- The Petri dishes must be autoclaved at a minimum of 120 °C before safe disposal.

Common errors to be aware of

Remind students of the technique for spreading the bacteria on the agar correctly to avoid not having enough bacteria or destroying the agar itself.

Supporting your students

Ask students to show you the pencil technique from Practical investigation 10.1 before allowing them to spread the loop onto the agar.

Challenging your students

Ask students to research and suggest why certain areas such as toilets and sinks could not be used for this investigation. Ask students to suggest reasons why the agar plates are incubated upside down.

Discussion points and scientific explanation

Build upon the discussion of Practical investigation 10.1 and reinforce the need for thorough safety precautions. Compare the differences in bacterial growth for different groups. Discuss why more bacteria were evident from certain areas. If you have taken a sample from a door handle or a device that is handled by many students, you should be able to observe larger colonies of growth due to the increased bacteria that may have been there initially.

Answers to workbook questions

1 Method:
   1 soap and water; disinfectant
   2 name
   3 Bunsen
   4 inoculating loop
   7 Petri; agar
   9 25
   10 workbench

2 Student's labelled drawing
3 Student's estimate
4 Student's actual count
5 Student answer denotes location of sample taken.
6 Student describes the size, number and spread of the bacteria in their Petri dish.
7 Students identify any of the safety precautions listed in 'Safety considerations' and provide a suitable explanation for them as outlined in the following list:
   - Wear safety spectacles at all times – protect eyes from possible infection.
   - Clean all equipment with disinfectant or similar solution, before and after the investigation – create a sterile environment at the beginning of the investigation and sterilise the area after investigation for the safety of other people.
   - Do not open the sealed Petri dish – dangerous pathogenic bacteria may escape and cause infections to people in the room.
   - Take care when setting up and using the Bunsen burner – avoid burning hands on the hot Bunsen, or causing injury/burns to other students.
   - Do not eat or drink in the laboratory – to avoid ingesting or inhaling any airborne pathogens that may be present.
   - Do not ingest or inhale the contents of the Petri dish or the bottle of nutrient broth – to avoid illness or infection from pathogens entering the body.
   - Cover all open wounds or cuts – to avoid illness or infection from pathogens entering the body.

Practical investigation 10.3
Effect of antibacterial mouthwashes on bacteria

Planning the investigation

In this investigation, students will compare the effectiveness of different antibacterial mouthwashes at destroying bacteria.

The safety precautions for this investigation should be drawn from your safety data sheets, local regulations and a microbiology safety guide for the materials and substances that you use.

This investigation will focus on the following assessment objectives:

- AO3.1 Demonstrate knowledge of how to safely use techniques, apparatus and materials (including following a sequence of instructions where appropriate)
- AO3.3 Make and record observations, measurements and estimates
- AO3.4 Interpret and evaluate experimental observations and data.

Setting up for the investigation

Make an advance purchase and storage of sterile agar plates from your local scientific supplier.
The method suggests that students clean their benches with disinfectant; however, this can be done by the technician before the lesson starts if possible.
If possible, prepare the paper discs in advance by using the circles taken from a hole punch and soaking them in three different antibacterial mouthwashes. An interesting extension task can be set by keeping the bottles handy with their prices displayed on the bottle. Equipment list per group of 2–4 students: disinfectant, safety spectacles, ruler, sterile agar plate, marker pen, microbe sample in nutrient broth, Bunsen burner and heat mat, inoculating loop, sticky tape, paper discs soaked in three different antibacterial mouthwashes and one soaked in water, access to an incubator at 25 °C. If you do not have access to an incubator, leave the dishes at room temperature in a safe place such as a lockable prep room.

Safety considerations

MH **HH**

You must consult your safety data sheets and guidelines as well as familiarise yourself with basic microbiology precautions relevant to the materials being used. Prior to carrying out the investigation, consult with the laboratory technician or teacher with relevant microbiological experience. Adhere to the following safety rules at all times:

- Wear safety spectacles.
- Do not open the Petri dish once it has been sealed.
- Do not eat or drink in the laboratory.
- Do not ingest or inhale any of the substances at any time.
- All open cuts or wounds must be covered.
- The Petri dishes must be autoclaved at a minimum of 120 °C before safe disposal.

Common errors to be aware of

Remind students of the technique for spreading the bacteria on the agar correctly to avoid either not having enough bacteria or destroying the agar itself. The paper discs must be placed onto the agar with space between them so that the clear zone can be identified and measured.

Supporting your students

Pair these students with more able students, or work closely with them to support the many different stages of the method. By working together as a group with an explanation of each stage, the method should be relatively straightforward to follow.

Challenging your students

Ask students to make predictions about which antibacterial mouthwash will be the most effective. They can use the bottle, the price of the mouthwash or even the ingredients listed on the bottle to make their decisions. Encourage students to discuss how accurate they were with their predictions. Students can then write a brief article about what they discovered, such as 'Does the price of a mouthwash indicate the quality?'

Discussion points and scientific explanation

Build upon the discussion of Practical investigations 10.1 and 10.2 to reinforce the need for thorough safety precautions. Compare the differences in bacterial growth for different groups. Discuss why different mouthwashes have different actions on the same bacteria. Do any of these students use these mouthwashes? Would they change their choice of mouthwash for another brand based on their investigation?

Answers to workbook questions

1. List of solutions used by the students
2. Students' drawings
3. Student answers match their Petri dish, units provided (mm)
4. Student answer
5. The antibacterial agent was the most effective and was able to destroy more of the bacteria.
6. They destroy bacteria in the mouth that may otherwise attack our teeth. This prevents tooth decay and gum disease.
7. To observe if it was the antibacterial mouthwash that had an effect on the bacterial growth, as a control for the investigation

Answers to exam-style questions

1. Disease-causing [1] organism [1]
2. Any three answers from the following: skin acts a physical barrier, mucus traps bacteria in the airways, hair in the nose traps microbes and prevents entry, stomach acid destroys bacteria before entering the body, blood clots seal wounds to prevent entry by pathogens
3. a Immunise [1] against specific pathogens [1]
   b Vaccine contains weakened or dead bacteria or viruses [1] that are recognised by lymphocytes which produce specific antibodies [1]. The lymphocytes produce memory cells [1] so if Jordan were to contract the 'real' viruses or come into contact with the bacteria, then his body would respond by making the correct antibodies to immediately fight the infection [1].

11 Respiration and gas exchange

Overview

In this chapter, students will investigate the action of respiration under different conditions, as well as how this affects the rate of gaseous exchange. This will provide them with the understanding of how aerobic respiration and anaerobic respiration benefit organisms.

Practical investigation 11.1 Germinating peas

Planning the investigation

In this investigation, students will investigate the temperature change of germinating peas and link this to the cause by respiration. Students will consider the role of a control as part of a reliable method.

This investigation will focus on the following assessment objectives:

- AO3.1 Demonstrate knowledge of how to safely use techniques, apparatus and materials (including following a sequence of instructions where appropriate)
- AO3.3 Make and record observations, measurements and estimates
- AO3.4 Interpret and evaluate experimental observations and data
- AO3.5 Evaluate methods and suggest possible improvements.

Setting up for the investigation

Advance purchase of vacuum/insulated flasks is required and dictates the amount of materials required to successfully complete the investigation. The temperature changes are relatively small in this investigation, which is why a vacuum flask works best. The smaller the better, as this then requires fewer peas to be bought.

Identify the quantity of peas required per flask and scale this up to purchase enough peas for the investigation.

Depending on your budget and storage capabilities, choose to have one experiment set up for the whole class or per small group.

Equipment list per group (1–50 students): vacuum flask × 2, peas soaked in water for 24 hours, cotton wool cut into strips suitable for filling the neck of the flask, thermometer × 2, boiled peas (this can be done as part of the student method or prepared in advance), dilute disinfectant. The peas require at least 48 hours for a temperature change but it is up to you to decide when the temperature is to be taken by students. This will depend on the teaching schedule and time between lessons.

Safety considerations

MH

The cleaning of the peas in the dilute disinfectant minimises the small risk of excessive bacteria or fungi growing on the peas.

Common errors to be aware of

Not diluting the peas with disinfectant may cause growth of bacteria or fungi and affect the temperature change. Students would not be able to say for sure whether the temperature change was due to the germinating peas or the growth of these microbes.

Supporting your students

Working in small groups offers support for struggling students who may have poor motor skills. Encourage students with excellent practical skills to be responsible for the lining of the cotton wool, the turning of the flask, and the supporting of the flask in an upside down position. Less able students can focus on producing a group table or recording the starting temperature.

Challenging your students

Direct students to collect data from other groups and calculate a class average for the investigation. Ask them to explain any anomalies that may be present.

Discussion points and scientific explanation

The boiled peas remained at the same temperature or increased by a degree or two at the most. These peas were the control in order to show that the germination of the other peas was most likely responsible for the increase in temperature. The germinating peas increased by 5–15 °C depending on the time allowed. Discuss how this is most likely due to respiration producing the heat energy but may also come from the heat energy produced in other reactions in the peas.

Answers to workbook questions

1. Student table
2. Student answers
3. Student answer is the difference between the two answers given in Question 2.
4. Heat energy released during respiration
5. To reduce heat loss to the surroundings
6. They act as a control.
7. To observe the temperature change and link this to the germination of the seed, so that any temperature change was not due to growth/respiration of bacteria/fungi
8. Collate results and calculate the mean

Practical investigation 11.2 Lung dissection

Planning the investigation

In this investigation, students will investigate the structure and function of a lung, and observe the differences before and after inflation. Students will suggest an investigation to measure the differences in the lungs before and after breathing.

This investigation will focus on the following assessment objectives:
- AO3.1 Demonstrate knowledge of how to safely use techniques, apparatus and materials (including following a sequence of instructions where appropriate)
- AO3.2 Plan experiments and investigations
- AO3.3 Make and record observations, measurements and estimates.

Setting up for the investigation

Advance purchase of a suitable lung is required. Consult with your local butcher to order suitable lungs that are available in your area. You will need to purchase as many as you require providing each group of 2–5 students with a lung each. Avoid the use of a pig's lung, as this may be culturally insensitive to some students. Prepare the dissection trays in advance or allow students to identify and select the equipment required as they see fit.

Equipment per group of 2–5 students: mammal lung, dissection tray, scalpel, dissection scissors, disposable gloves, safety spectacles, clear plastic tubing.

Safety considerations

MH

Wear safety spectacles and gloves at all times. Wash hands afterwards, no food or drink allowed, and take care when using sharp tools. Clean all surfaces with disinfectant after the investigation has finished. Cover the lungs with a clear plastic bag prior to inflation to avoid bacteria escaping from the cut surfaces of the lung.

Common errors to be aware of

Students may accidentally suck through the tubing instead of blowing. It is advised that the action is clearly demonstrated by the teacher without the lung attached. Students can then try doing so in the same manner before attaching the tube to the lungs.

Supporting your students

Pair these students with a more able student and encourage the struggling student to be the one who feels the inflated lung. Student can refer to diagrams of the lung in order to try to match the different structures to what they can see and feel.

Challenging your students

Allow students to measure the lung before and after inflation. They can use this information to calculate the percentage change in size. As an alternative to Question 3, students can plan a method to measure and compare the surface area of the lung before and after inflation. Students could also produce a table linking the structures of the lungs to their respective functions.

Discussion points and scientific explanation

Discuss the parts of the lung in detail, following the route that an oxygen molecule might take from mouth to alveoli. Explain the different structures and their function, involving the students in the discussion. Students should be able to explain that, as the lung increases in size, it increases the surface area and allows more gaseous exchange to take place.

Answers to workbook questions

1. Student labelled diagram
2. **a** The lung floated.
   b The density of the lung is less than the density of water, due to the amount of air and air space in the lung, even when deflated.
3. Clean with disinfectant to remove harmful bacteria.
4. Student answer should plan a method to measure the width/length/size of the lung before inflation and after inflation. Student should suggest working out the difference, or percentage change in size.

Practical investigation 11.3
Effect of exercise on breathing rate

Planning the investigation

In this investigation, students will carry out an investigation to measure their breathing rate after several minutes of exercise. Students will use their results to plot their results as a line graph.
This investigation will focus on the following assessment objectives:
- AO3.1 Demonstrate knowledge of how to safely use techniques, apparatus and materials (including following a sequence of instructions where appropriate)
- AO3.2 Plan experiments and investigations
- AO3.3 Make and record observations, measurements and estimates
- AO3.5 Evaluate methods and suggest possible improvements.

Setting up for the investigation

Identify a suitable area for exercise. This could be an outdoor area depending on the weather, or a large room or hall if you have one available. Students can do on-the-spot exercises such as star-jumps, squats, walking- and jogging if space is limited.
Equipment required per group of 2–4 students: space, stopwatch.

Safety considerations

Awareness of the area should be assessed and communicated to the students. Explain to students that they should stop exercising if they feel unwell or dizzy.

Common errors to be aware of

Some students might not allow their breathing rate to return to the resting rate. Encourage students to allow as much time as possible for this in their planning. They can also rotate the roles of the group so that, when one person is resting, another is recording or exercising.

Supporting your students

Remind students of the steps for planning a method, making a table and plotting a graph. Display the variables of the investigation and discuss where they will be placed on the table and the graph.

Challenging your students

Ask students to calculate the percentage increase in their breathing rate and discuss the reasons behind the different changes for each person.

Discussion points and scientific explanation

Compare the results for different students and discuss the need for increased oxygen during and after exercise. The body requires more oxygen in order to do more respiration to release the energy required for the increase in activity. Ask students to suggest reasons why some students had a much greater increase in breathing rate during and after exercise.

> **Answers to workbook questions**
>
> 1. Student method suggests a sensible method to record the breathing rate by measuring the number of breaths per minute at different time intervals.
> 2. Awareness of safe area and space, student to monitor well-being of those exercising
> 3. Student table has the breathing rates (breaths per minute) recorded at the stated time intervals.
> 4. Student graph plots the data for time (minutes) on the x-axis and breathing rate (breaths per minute) on the y-axis.
> 5. Breathing rate increases as time exercising increases.
> 6. The body requires more oxygen for respiration so breathing rate increases to increase the amount of oxygen being taken in.

Practical investigation 11.4
Repaying the oxygen debt

Planning the investigation

In this investigation, students will independently plan an investigation to observe the time taken to repay the oxygen debt after exercise. Students will link this to the recovery period and the time taken to break down lactic acid produced during the exercise. This investigation is an ideal opportunity for students to practise their planning skills but the method will only cover the rate of breathing. It will not account for VO_2 Max, or the volume of oxygen uptake, unless you have access to the appropriate breathing apparatus. If this is the case, then this apparatus may be easily used to measure the depth AND rate of breathing.

This investigation will focus on the following assessment objectives:
- AO3.1 Demonstrate knowledge of how to safely use techniques, apparatus and materials (including following a sequence of instructions where appropriate)
- AO3.2 Plan experiments and investigations
- AO3.3 Make and record observations, measurements and estimates
- AO3.4 Interpret and evaluate experimental observations and data
- AO3.5 Evaluate methods and suggest possible improvements.

Setting up for the investigation

Find and clear a suitable area for exercise and provide a stopwatch for each group of students.

Safety considerations

Students must clear all tripping hazards. Provide water or water breaks for exercising students and do not exercise outside in extreme temperatures. Students must stop exercising immediately if they are feeling unwell or dizzy.

Common errors to be aware of

Students that measure their own breathing rate may subconsciously alter their breathing rate as they are thinking about it. Students should organise their groups to observe and record each other's breathing rates.

Supporting your students

Provide, or plan together, a detailed method before going out to the exercise area.

Challenging your students

Students may repeat the investigation measuring their heart rate or pulse. A comparison of data between the breathing rate and the heart rate can be plotted on a graph.

Discussion points and scientific explanation

Breathing rate increases during exercise as our muscles require more energy. The increase in energy demand is met by an increase in respiration, which requires more oxygen. More oxygen must be taken in by breathing deeper and faster. It is possible that some students may not observe an increase in breathing rate (some may even observe a decrease). If this is the case, then the discussion should focus on why this has happened. It is likely that the depth of breathing (i.e. the amount of oxygen taken in) will be much greater and thus there is less need for faster breaths. A suitable extension activity or question would be for students to research and plan how they could measure the volume of oxygen taken in by each person.

Lactic acid is a product of anaerobic respiration and its accumulation causes a pH change. It is produced during hard exercise and must subsequently be removed, as it would be dangerous to the body if the body was not capable of preventing these levels being reached. Oxygen is required to convert the lactic acid to carbon dioxide and water; therefore the breathing rate must remain high enough to meet this demand for oxygen. The time taken to break down the lactic acid in the liver is the recovery period.

Answers to workbook questions

1. Student method should be similar in style and structure to the method in Practical investigation 11.3. The student should state how the breathing rate is achieved and measured, the type and length of exercise, and how the breathing rate will be measured every minute after exercise until the rate returns to the resting rate.
2. The table should include the following headings for each student tested: Breathing rate at rest / breaths per minute, Breathing rate for every minute after exercise (this may go up to 5 to 10 minutes depending on activity and student fitness).
3. Student graph should match data from the table with Time/minutes on the x-axis, and Breathing rate/breaths per minute on the y-axis. The graph should be drawn to an appropriate scale and the plots joined point-to-point with a ruler.
4. Student describes the increase in maximum breathing rate during exercise, before describing the gradual decrease in breathing rate after the exercise has finished.
5. To provide oxygen required to break down the lactic acid produced during exercise
6. Lactic acid
7. The time taken for the body to break down the lactic acid and return to the normal breathing rate
8. For safety reasons to ensure that the volunteers do not overdo the exercise and become ill

Answers to exam-style questions

1.
   a Person is breathing at a normal rate [1]
   b Person takes a deep breath [1]
   c Residual volume [1]
   d Working out shown, $3\,dm^3 - 2.5\,dm^3$ [1] shown, answer = 0.5 [1] dm^3 [1]
   e Working out shown, $5.5\,dm^3 - 2.5\,dm^3$ [1] shown, answer = 3 [1] dm^3 [1]
2. Student answer includes the following points: Breathing rate is 'normal' before running [1]. Breathing rate increases gradually as she runs [1]. Breathing rate reaches a maximum level [1]. Breathing rate increases to take in more oxygen for respiration [1]. Allow student credit if their answer includes reference to breathing deeper as part of the increase in oxygen uptake.
3.
   a Chemical reaction [1] in cells that use oxygen [1] to break down nutrient molecules [1] to release energy [1]
   b $C_6H_{12}O_6 + 6O_2 \rightarrow 6CO_2 + 6H_2O$ [3]
4.
   a Chemical reaction [1] in cells that break down nutrient molecules [1] to release energy [1] without using oxygen [1]
   b $C_6H_{12}O_6 \rightarrow 2C_2H_5OH + 2CO_2$ [3]
5.
   a Student graph shows breathing rate on the y-axis [1] against time on the x-axis [1], with appropriate units [1]. Curve sketched showing breathing rate at rest for 3 minutes, before rising to a peak at 10 minutes, falling back to the resting rate at 14 minutes [1].
   b Sam's breathing rate increases gradually from 3 to 30 minutes [1], before decreasing back to the resting rate at 14 minutes [1].
   c Sam's muscles had produced lactic acid during exercise [1], which requires oxygen to break down into carbon dioxide and water [1]. Breathing rate remains high [1] until this oxygen debt [1] has been paid off.

12 Excretion

Overview

In this chapter, students will investigate how the body removes some waste products such as carbon dioxide, as well as those excreted in faeces and urine. The removal of carbon dioxide continues on from what they learnt in Chapter 11.

Practical investigation 12.1
Kidney dissection

Planning the investigation

In this investigation, students will dissect a mammal kidney and observe the internal structures of the kidney. They will reinforce their drawing and magnification skills from earlier units.

This investigation will focus on the following assessment objectives:
- AO3.1 Demonstrate knowledge of how to safely use techniques, apparatus and materials (including following a sequence of instructions where appropriate)
- AO3.3 Make and record observations, measurements and estimates
- AO3.4 Interpret and evaluate experimental observations and data
- AO3.5 Evaluate methods and suggest possible improvements.

Setting up for the investigation

Purchase a kidney in advance from your local butchers or supermarket. Sheep kidneys and lamb kidneys are suitable but anything of similar size will suffice. Avoid the use of a pig's kidneys as this may be culturally insensitive to some students.

Equipment list per group of 2–4 students: mammal kidney, dissection tray or board, dissection scissors and scalpel, latex gloves, ruler, safety spectacles, weighing balance.

Safety considerations

Students will plan their own safety precautions for this investigation. Check these before beginning as well as providing your own safety brief. Wear safety spectacles and gloves, dispose of kidneys safely, take care using the dissection tools, and wash hands afterwards. All dissection equipment must be cleaned with disinfectant or similar by you or the technicians.

Common errors to be aware of

Students may not make a suitable cut through the middle of the kidney. Some students may cut all of the way through. A clear demonstration by the teacher should be enough to guide the students on the ideal cut. Taking the time to identify where and how the cut will be made is crucial for the teacher and student – just like a surgeon, you only get one chance to make the correct incision!

Supporting your students

Observe closely and guide them individually when making the incisions.

Challenging your students

Direct students to add an outline of the functions of the internal structures on their labelled drawing.

Discussion points and scientific explanation

Point out the main areas that can be viewed inside the kidney such as the renal pelvis, the renal artery and vein, the medulla and the cortex. Discuss the role of each part of the kidney and the pathway that blood may take through the kidney.

> **Answers to workbook questions**
>
> 1. Student safety plan
> 2. Student records
> 3. Correct formula shown (magnification = image size/actual size), student answer greater than 1
> 4. Student correctly identifies one of the internal structures and links this to the appropriate function.
> 5. Any sensible suggestion such as using a microscope or magnification instrument of some kind

Practical investigation 12.2
Expired and inspired air

Planning the investigation

In this investigation, students will investigate the difference in composition between expired air and inspired air. They will observe and explain the differences in levels of oxygen, carbon dioxide, water vapour and temperature.

This investigation will focus on the following assessment objectives:
- AO3.1 Demonstrate knowledge of how to safely use techniques, apparatus and materials (including following a sequence of instructions where appropriate)
- AO3.3 Make and record observations, measurements and estimates
- AO3.4 Interpret and evaluate experimental observations and data.

Setting up for the investigation

Equipment per group of 2-4 students: thermometer, limewater, boiling tube, mirror, cobalt chloride paper, wooden splint, matches, plastic straws, safety spectacles, paper towels.

Safety considerations

MH

Wear safety spectacles and ensure that students do not accidentally ingest the limewater by accidentally sucking the solution up through the straw. Demonstration of breathing in before blowing gently into the straw is useful.

Common errors to be aware of

Students may blow too hard into the limewater causing it to explode out of the boiling tube. Use of the paper towel to create a makeshift lid will prevent limewater from spilling out.

Supporting your students

Demonstrate each of the tests to struggling students before allowing them to follow your lead. Other students will be able to follow the method independently.

Challenging your students

Direct students to produce a table of inspired air versus expired air, showing the composition of each type of air.

Discussion points and scientific explanation

Discuss the results with students, encouraging them to explain why each result was produced. The temperature of expired air increased because air is warmed as it passes through the respiratory system. The water vapour that escapes as part of expired air will test positive for water under cobalt chloride paper. The limewater turned cloudy/milky as a positive test for carbon dioxide. Oxygen relights a glowing splint; although this may not have happened in this instance, the splint will have at least glowed more brightly than originally.

> **Answers to workbook questions**
>
> 1. (**a**–**d**) Student observations
> 2. **a** Temperature of air increases as it passes through the respiratory system.
>   **b** Water vapour will test positive for water under cobalt chloride paper and turns the paper blue.
>   **c** Limewater turns cloudy/milky in the presence of carbon dioxide.
>   **d** Oxygen relights a glowing splint.

Answers to exam-style questions

1   a   Removal of nitrogen-containing part of amino acids [1] to form urea [1]
    b   D, A, E, C, B [5]
2   Maximum of three from: high success rate [1], easier than attending hospital for regular dialysis [1], overall financial cost reduced compared to dialysis [1], frees up doctors and nurses from administering regular dialysis [1], or any other sensible suggestion

13 Coordination and response

> **Overview**
>
> In this chapter, students will investigate how the human body reacts to different stimuli and link this to the purpose of voluntary and involuntary reactions.

Practical investigation 13.1 Measuring reaction times

Planning the investigation

In this investigation, students will investigate their reaction times and plan a suitable table for recording their results.

This investigation will focus on the following assessment objectives:

- AO3.1 Demonstrate knowledge of how to safely use techniques, apparatus and materials (including following a sequence of instructions where appropriate)
- AO3.3 Make and record observations, measurements and estimates
- AO3.5 Evaluate methods and suggest possible improvements.

Setting up for the investigation

Provide students with half-metre rulers. Metre rulers can be used but are heavy and cumbersome. Students may also use their own 30 cm rulers but will occasionally not react quickly enough and miss the ruler completely. Students will need to be able to rest their arms on benches or tables.

Safety considerations

Do not use old wooden rulers that may give students a cut or splinters.

Common errors to be aware of

Students anticipating the release of the ruler rather than relying on their reaction times when they see the ruler being released. Tell students to vary the length of time before they release the ruler to minimise the anticipation.

Supporting your students

Plan the results table in a small group before beginning the method. Students then just have to enter their data for the investigation.

Challenging your students

Direct students to compare their reaction times to those that they can research on the internet. Direct them to find out the reaction times of a sprinter and a Formula 1 driver and compare.

Discussion points and scientific explanation

Discuss the role of anticipation in the results and how this can be avoided. Most students will improve their reaction times with practice. Use this to lead into a discussion about possible exercises for sportsmen or people that require very quick reaction times.

Answers to workbook questions

1. Student table
2. Student answer
3. Yes, a combination of anticipation, and the body reacting faster to the stimulus
4. Student prediction
5. Outline of method similar to original method but with the ruler touching the hand before release
6. Accept suggestions such as use of data loggers to remove human error or the use of video footage to capture and check exact distances.
7. Suggestion of method involving blindfold and reaction to a sound stimulus, such as a bell attached to the ruler

Practical investigation 13.2
Sensitivity test

Planning the investigation

In this investigation, students will test the sensitivity of different parts of their body and link this to the presence of receptors at or near the surface.

This investigation will focus on the following assessment objectives:

- AO3.1 Demonstrate knowledge of how to safely use techniques, apparatus and materials (including following a sequence of instructions where appropriate)
- AO3.2 Plan experiments and investigations
- AO3.3 Make and record observations, measurements and estimates
- AO3.5 Evaluate methods and suggest possible improvements.

Setting up for the investigation

Little set-up required other than purchase of wire or paper clips.

Safety considerations

Warn students not to press too firmly with the wire, or test exposed areas such as the mouth, eyes, ears and nose.

Common errors to be aware of

Students do not keep the distance between the two pins equal for every test. Students can eliminate this by measuring the 5 mm gap each time they use their piece of bent wire to touch the skin of their partner.

Supporting your students

Pair students with a more able student who can take the lead with this activity. Ask the more able student to explain each step.

Challenging your students

Direct students to try moving the two ends of the wire closer together and retest the areas that could detect the two pins. Discuss how this can be used to potentially identify the most sensitive part of the body.

Discussion points and scientific explanation

The areas of the body that could detect the two pins are more sensitive than the areas that could detect only one pin. This is because there are more nerve endings in these areas, which are then able to detect the two pins as separate stimuli.

Answers to workbook questions

1. Student table
2. Student answer
3. Student answer
4. More receptors in those parts of the body so the two pins could be detected separately
5. Method outlined with suggestion such as student being blindfolded and not being told which part of the body would be tested

Practical investigation 13.3
Human responses

Planning the investigation

In this investigation, students will carry out an investigation that will cause involuntary reactions to a range of stimuli. Students will link these responses to their relevant use in the body.

This investigation will focus on the following assessment objectives:
- AO3.1 Demonstrate knowledge of how to safely use techniques, apparatus and materials (including following a sequence of instructions where appropriate)
- AO3.3 Make and record observations, measurements and estimates.

Setting up for the investigation

Obtain enough half-metre rulers or similar implement suitable for the investigation. If you have one, use a doctor's hammer (similar to what a doctor uses for the knee reflex test). Equipment list per group of 2–4 students: torch, half-metre ruler, chair, table or bench.

Safety considerations

Students must take care when climbing on or off the tables and chairs. Ensure that students do not tap too hard by demonstrating this clearly as part of the teacher demonstration. Warn students not to press too firmly on their eyelid otherwise they might cause temporary pain or discomfort to the eye.

Common errors to be aware of

Students may not tap the correct area for the knee and/or heel reflex. Demonstrate where this should be done. If you have not done this before, try it out on one of your colleagues in advance so that you know exactly how to demonstrate to the students.

Supporting your students

Draw a diagram for each of the five tests to provide a visual aid to the method.

Challenging your students

Ask students to research another test for involuntary actions and try them out on each other.

Discussion points and scientific explanation

Discuss the uses behind each of the reactions and how they are all involuntary. The actions happen without having to think about them. The size of the pupil changes to allow less light into the eye when a torch is shone in to protect the eye from damage from excessive light. The reaction of the knee and the heel are crucial to the action of walking. Discuss how difficult walking would be if students had to think about bending the different parts of the leg and foot every time a step was taken. The eye reacts to sudden movement by blinking to prevent foreign objects from entering the eye. The eyelid pressure test caused the messages sent to the brain to be distorted, causing temporary double vision. This has no use to the human body but demonstrates how the eyes and brain work together to form the image that you see.

Answers to workbook questions

1. Completed table
2. The pupil becomes smaller to prevent excessive light damaging the eyes; the reaction of the heel and the knee are involuntary actions useful for walking; and the eyelids close and open quickly to prevent foreign objects from entering the eye. The double vision occurs because the pressure on the eyelid distorts the signal sent to the brain.
3. Each person is different and reacts in different ways; some students might not have the stimuli carried out in the same way. For example, some student might have tapped the heel or the knee in the wrong place.

Answers to exam-style questions

1. a Student draws growth of plant navigating away around the maze [1] to reach the top of the box where there is light [1].
   b Positive [1] phototropism [1]
   c Auxin [1]
   d Response of a plant that grows away [1] from light [1]
2. a X is the retina [1], Y is the optic nerve [1], and Z is the cornea [1].
   b Transfers electrical impulses [1] from receptors to the brain in order to form an image [1]
   c Excessive light causes damage [1] to the retina [1]

14 Homeostasis

Overview

In this chapter, students will investigate how the human body is affected by its external environment and how it responds in order to maintain a constant internal environment.

Practical investigation 14.1 Controlling body temperature

Planning the investigation

In this investigation, students will investigate how insulation prevents heat energy escaping from a system. This will represent the effectiveness of the insulating layer of air trapped by erect hairs during thermoregulation in the human body.

This investigation will focus on the following assessment objectives:

- AO3.1 Demonstrate knowledge of how to safely use techniques, apparatus and materials (including following a sequence of instructions where appropriate)
- AO3.3 Make and record observations, measurements and estimates
- AO3.4 Interpret and evaluate experimental observations and data.

Setting up for the investigation

Select several different materials for use as insulating materials. Select a wide range to allow students to investigate many different materials across the different groups. Do not cut the material to size – this should come from the student as they aim for a reliable method. Use materials such as paper, magazines, bubble wrap, aluminium foil, thin cloth, nylon, cotton wool, thick cloth or any other similar materials suitable for the investigation.

Purchase polystyrene cups in advance, or use 250 ml beakers. The number of cups or beakers being tested at one time varies depending on the number of thermometers, sources of hot water, and students in the group. Students can either test all of the materials at the same time, or test them one at a time as suggested in the method in the student workbook.

Equipment per group of 2–4 students: polystyrene cups (or beakers) × 4, insulating materials, elastic bands × 4, thermometer, hot water, stopwatch.

Safety considerations

Students must take care when using the hot water.

Common errors to be aware of

Students often leave the water uncovered or cooling for too long before insulating and taking the temperature. Preparation of the material, teamwork and practice will increase the chances of obtaining reliable data.

Supporting your students

Carry out the investigation with fewer materials that will give obvious results. This could be one cup without material, one covered in paper, and one covered in aluminium foil.

Challenging your students

Direct students to improve reliability by repeating the investigation more than once and calculate the mean temperature change of the materials. Students can practise their graphing skills by plotting a bar chart to show the different changes in temperature of the materials.

Discussion points and scientific explanation

Discuss how an insulating layer can prevent heat energy escaping from a system. The beaker or cup represents the body and the material represents the insulating layer of air, fat and clothing that a cold person uses to maintain a constant internal temperature.

Answers to workbook questions

1. Student table
2. Student calculations
3. Student calculations
4. Student describes, using data from the table, how the temperature decreases for all of the tests and identifies which materials caused the most/least change.
5. The material prevents the heat energy from escaping and keeps it trapped inside the beaker or cup.
6. The hairs on the outside of the skin become erect, causing an insulating layer of air to be trapped. This, along with a layer of fat and any clothes that the person may be wearing, minimises the heat energy escaping from the body.
7. To make the investigation (more) reliable
8. The same volume of water at the same starting temperature could be used.
9. This is a control to see that the materials did have an effect on the rate of cooling of the beakers/cups.

Practical investigation 14.2 Effect of body size on cooling rate

Planning the investigation

In this investigation, students will investigate how the size of a body affects the rate of cooling by observing the temperature change of a body of water in different sized beakers.

This investigation will focus on the following assessment objectives:
- AO3.1 Demonstrate knowledge of how to safely use techniques, apparatus and materials (including following a sequence of instructions where appropriate)
- AO3.3 Make and record observations, measurements and estimates
- AO3.4 Interpret and evaluate experimental observations and data.

Setting up for the investigation

Boil water in a kettle or provide multiple Bunsen burners and relevant equipment for students to heat their own water. Students may carry out the investigation for each beaker at the same time if you have enough thermometers.

Equipment list per group of 2–4 students: glass beakers (50 ml, 100 ml and 250 ml), thermometer, stopwatch, hot water.

Safety considerations

Students should take care when using hot water.

Common errors to be aware of

Students may not be able to judge the halfway point for pouring the water into the beakers. Allow students to fill the water to another obvious point, such as the top of the beaker. Reinforce need for safe handling and placement of beaker if this is chosen.

Supporting your students

Support students by planning and drawing the results table together prior to beginning the method. Demonstrating the investigation ahead of theirs can help students carry out their own.

Challenging your students

Direct students to research large and small animals to find out if their results represent what happens in real life. Students should focus on the ratio of surface area to volume.

Discussion points and scientific explanation

Animals lose heat quickly if they have a large surface area compared to their volume. An extremely large animal, such as an elephant, will lose heat energy slowly to their surroundings because they have a small surface area compared to their volume.

Answers to workbook questions

1. Student table
2. Student graph
3. Student answer describes the different rates of temperature loss as shown in their graph.
4. Any from volume of water used, same type of beaker, same thermometer, same surrounding temperature and conditions for each beaker, same starting temperature, same stopwatch
5. Any sensible answer for Question 2, to improve the reliability of the investigation

Practical investigation 14.3 Evaporation rates from the skin

Planning the investigation

In this investigation, students will investigate how the rate of evaporation can affect the cooling effect observed on the skin. The student will link this to thermoregulation as part of homeostasis.

This investigation will focus on the following assessment objectives:
- AO3.1 Demonstrate knowledge of how to safely use techniques, apparatus and materials (including following a sequence of instructions where appropriate)
- AO3.3 Make and record observations, measurements and estimates
- AO3.4 Interpret and evaluate experimental observations and data
- AO3.5 Evaluate methods and suggest possible improvements.

Setting up for the investigation

Prepare small quantities of water and acetone for student use. If acetone is not available, perfume, aftershave or some form of alcohol can be used instead. Data loggers can be used instead of thermometers if available. Equipment list per group of 2–4 students: thermometer × 3, test-tube rack or boss clamps × 3 and stand, acetone, water, cotton wool balls, safety spectacles.

Safety considerations

Take care when handling acetone and wear safety spectacles. Consult your safety data sheets if using a substance other than acetone.

Common errors to be aware of

Students may soak the cotton wool balls before they place them on the thermometer, so that some heat loss might not be fully observed. Teacher demonstration and reminder should avoid this error.

Supporting your students

Small groups of students will benefit from teacher support if struggling to get the cotton wool soaked while on the thermometer. Use of a pipette or forceps to soak the cotton wall ball before placing onto the thermometer might be easier (even if some reliability is lost, the results should still lend themselves to a suitable conclusion).

Challenging your students

Direct students to plan the same investigation but to use a data logger. Ask students to explain how this will improve the accuracy of the investigation. Students can suggest solutions other than acetone and water to be tested.

Discussion points and scientific explanation

The evaporation of the water and the acetone removes heat energy from the cotton wool ball, resulting in a drop in temperature. This reflects what happens during sweating as the evaporation of the water in sweat has a cooling effect on the skin. The cooling effect is part of the body's thermoregulatory response to an increase in internal body temperature.

Answers to workbook questions

1. Student table
2. Student answer describes the different changes, using data from the table to support that the cotton wool ball soaked in acetone suffered the greatest drop in temperature, followed by the cotton wool ball soaked in water. The dry cotton wool ball showed no (or negligible) change in temperature.
3. The alcohol in the aftershave evaporates quickly, having a pronounced cooling effect on the skin.
4. As the water evaporates, the heat energy leaves the skin and lowers the internal body temperature. This results in feeling cooler.
5. Student table is similar to the table in the investigation but shows an extra column for repeat readings and the calculation of the mean change in temperature. All headings and units are included.
6. Student can suggest that more readings can be taken, the change in temperature can be monitored constantly rather than at specific time points, and the overall accuracy of the investigation improves.
7. To compare what happens when no solution is added to the cotton wool

Answers to exam-style questions

1. The maintenance of a constant internal environment [1]
2. a Line graph [1] with sensible scale covering at least half of the paper [1], correctly plotted [1], axes labelled with units [1], points joined together by a line going through each of the points [1]
   b As time passes [1], body temperature decreases [1].
   c Shivering (rapid contraction and relaxation of the muscles) [1], hairs at the skin become erect [1] and trap an insulating layer of air [1], description of vasoconstriction [1] to reduce blood flow and heat loss from the blood [1]

15 Drugs

> **Overview**
>
> In this chapter, students will investigate the effect of caffeine on their reaction time and relate that to how chemical substances (drugs) can affect the human body. They will also investigate the effectiveness of different types of antibiotics, drugs used to tackle bacterial infections in animals including humans.

Practical investigation 15.1
Effect of caffeine on reaction times

Planning the investigation

In this investigation, students will compare data from Practical investigation 13.1 with data gathered testing the effect of caffeine on their reaction times.
This investigation will focus on the following assessment objectives:
- AO3.1 Demonstrate knowledge of how to safely use techniques, apparatus and materials (including following a sequence of instructions where appropriate)
- AO3.2 Plan experiments and investigations
- AO3.3 Make and record observations, measurements and estimates
- AO3.4 Interpret and evaluate experimental observations and data
- AO3.5 Evaluate methods and suggest possible improvements.

Setting up for the investigation

Equipment required for each group: half-metre ruler, caffeinated drinks, stopwatch, table or bench to rest arm on.
Prepare quantities of a caffeinated drink such as a can of cola or energy drink. Students should consume no more than one can in that day – warn students not to drink before the lesson, as this is a health hazard and may affect results. It is best to use cans or bottles of drinks as the quantity of caffeine can be checked on the side of the product. Using coffee requires more skill and carefully planned concentrations to be made up. Request permission from parents prior to the lesson and link this to the ethical or potential health concerns of drinking caffeine. Students who are not permitted to drink caffeine can still play an important role in the investigation.

Safety considerations

Consult the safety sheets for caffeine as a matter of good practice. There are many side-effects to continuous and excessive consumption of caffeine but the amount consumed in this investigation will be safe. Permission from parents will identify any students who should not drink caffeine. Warn students not to consume caffeine before or after the lesson, and share the health reasons why.

Common errors to be aware of

Students may not adhere to the 15-minute wait time. By doing the investigation at the same time as a group, this allows you to ensure that students have waited the appropriate time before testing their reactions. This is less of an error and more of an observation; some students drink caffeine and deliberately act as if they are affected by being silly. This behaviour, although rare, can be avoided by a reminder of laboratory safety expectations or use of a double-blind test. Carry out a double-blind test by preparing caffeinated and decaffeinated drinks in advance and providing them in cups marked by a code that only you know.

Supporting your students

Working as a group provides enough support for this straightforward investigation. Calculation of the averages and conversion of the reaction times can be done together as a class to support students with weaker mathematical skills.

Challenging your students

Direct students to research different caffeinated drinks and the amount of caffeine in them. Get them to research their favourite drinks or coffee shops and see how much caffeine is reported to be in their favourite drinks. They will be surprised and shocked at some of the figures. Ask the students to present back to the rest of the class with their findings.

Discussion points and scientific explanation

Discuss the range of reliability issues and variables of the investigation. Each person is different and may respond differently to the caffeine. Discuss a suitable method for minimising this variation, such as using the data from the class and calculating the mean reaction time before and after consuming caffeine. Caffeine is a stimulant and improves reaction times. In theory, all, or most, students should have improved their reaction times after drinking caffeine as their body and senses were on a higher level of alertness to respond to the stimulus of the falling ruler.

Answers to workbook questions

1. Student reaction time
2. Student table
3. Student calculations
4. Student compares the reaction time before and after drinking the caffeine.
5. Student links the effect to caffeine being a stimulant and the higher level of alertness meaning that the body can respond to stimuli much faster.
6. To allow enough time for the caffeine to have an effect on the body.
7.
    a. Student suggests a suitable method, including how the variable being changed (the caffeine) is kept hidden from participants of the test.
    b. To improve reliability of the method, to avoid bias by the person being tested
8. Student discusses whether students should be tested with drugs, even a legal and relatively harmless one such as caffeine. Student links the potential health effects of excessive caffeine use to poor health, diabetes, insomnia, illness, sore stomach, tooth decay.

Practical investigation 15.2
Effect of antibiotics on bacteria

Planning the investigation

In this investigation, students will compare the effectiveness of different antibiotics at destroying bacteria. The safety precautions for this investigation should be drawn from your safety data sheets, local regulations and a microbiology safety guide for the materials and substances that you use.

This investigation will focus on the following assessment objectives:

- AO3.1 Demonstrate knowledge of how to safely use techniques, apparatus and materials (including following a sequence of instructions where appropriate)
- AO3.3 Make and record observations, measurements and estimates

- AO3.4 Interpret and evaluate experimental observations and data.

Setting up for the investigation

Make an advance purchase of sterile agar plates from your local scientific supplier.

The method suggests that students clean their benches with disinfectant but this can be done by your technician before the lesson starts if possible.

Purchase a set of antibiotic discs (sometimes known as an antibiotic mast ring) from your local science supplier. These will often come in packs of six or eight and a key will be provided to identify the different codes for the different antibiotics. The mast rings can be placed in as a whole set or (as the workbook suggests) be cut up and used separately. It is recommended to use the full ring but you will need to direct the students to amend their method to account for this if you choose to do so. If you cannot get these pre-prepared packs, make up antibiotic solutions in advance and soak blotting paper rings or hole punch discs in them.

Equipment list per group of 2–4 students: disinfectant, safety spectacles, sterile agar plate, marker pen, Bunsen burner and heat mat, microbe sample in nutrient broth, inoculating loop, sticky tape, antibiotic discs, access to an incubator at 25 °C. If you do not have access to an incubator, leave the dishes at room temperature in a safe place such as a lockable prep room.

Safety considerations

HH

You must consult your safety data sheets and guidelines as well as familiarising yourself with basic microbiology precautions relevant to the materials being used. Consult in advance with your laboratory technician or teacher with the relevant experience in microbiological techniques before carrying out the investigation. Adhere to the following safety rules at all times:

- Wear safety spectacles.
- Do not open the Petri dish once it has been sealed.
- Do not eat or drink in the laboratory.
- Do not ingest or inhale any of the substances at any time.
- All open cuts or wounds must be covered.
- The Petri dishes must be autoclaved at a minimum of 120 °C before safe disposal.

Common errors to be aware of

Remind students of the technique for spreading the bacteria on the agar correctly to avoid not having enough bacteria or destroying the agar itself. The paper discs must be placed onto the agar with space between them so that the clear zone can be identified and measured; if they are joined together, simply cut them with a pair of scissors to produce three separate discs.

Supporting your students

Pair these students with more able students, or work closely with them to support the many different stages of the method. By working together as a group with an explanation of each stage, the method should be relatively straightforward to follow.

Challenging your students

Direct students to design the box that these antibiotics might be sold in. They can add keywords, information, and how antibiotics protect against disease to inform the buyer why they might need these antibiotics. Students can research the antibiotics used in the investigation to support their design.

Discussion points and scientific explanation

Build upon the discussion of Practical investigations 10.1 and 10.2 to reinforce the need for thorough safety precautions. Compare the differences in bacterial growth for different groups. Discuss why different antibiotics produced different results; link to the time taken for the mechanism of the antibiotics to take effect.

Answers to workbook questions

1. Names of antibiotics
2. Student drawings and labels
3. Student answers match their Petri dish, units provided (mm)
4. Student answer
5. The antibiotic destroyed more of the bacteria than the other two antibiotics.
6. Excessive growth, dangerous pathogens released, microbes could be inhaled/ingested/transferred, the next class might get bacteria on them, eating or drinking may cause ingestion of airborne pathogens or microbes, any other sensible answer related to the safety precautions stated in the workbook

Answers to exam-style questions

1.
   a. Antibiotic Y [1]
   b. Antibiotic X destroyed some bacteria [1] but not as many as antibiotic Y, which destroyed more bacteria. [1].
   c. Disc Z [1]
   d. Bacteria were not destroyed around disc Z [1].
   e. To compare the effect of the antibiotic, to confirm that the antibiotic was the main factor/variable that caused the bacteria to be destroyed [1]
2. Cilia in the airway are damaged and/or destroyed [1]. Damaged cilia are not able to remove mucus from the airways [1]. Goblet cells produce extra mucus [1] that cannot be removed by the paralysed cilia [1]. The mucus moves into the lungs and remains there [1].
3. Chemicals or substances that cause cancer/cells to divide over and over again [1]

16 Reproduction in plants

Overview

In this chapter, students will investigate the structure of a flowering plant and describe some of the parts that they can see. They will investigate the requirements for successful germination of cress seeds and explore the temperature dependence of plant growth. The close link between plant reproduction and the conditions required for germination is reviewed in this chapter.

Practical investigation 16.1 Structure of a flower

Planning the investigation

In this investigation, students will examine, dissect, measure and record the structure of a flower from an insect-pollinated flowering plant. This chapter covers the reproduction, fertilisation and germination of plants, and the practical investigation of germination is covered here. The environmental conditions have an important role to play in the overall process of the reproduction and growth of plants. The link between reproduction and germination must be established as part of these investigations.

This investigation will focus on the following assessment objectives:

- AO3.1 Demonstrate knowledge of how to safely use techniques, apparatus and materials (including following a sequence of instructions where appropriate)
- AO3.2 Plan experiments and investigations
- AO3.3 Make and record observations, measurements and estimates.

Setting up for the investigation

Advance purchase of flowering plants required; any plant with flowers in bloom would suffice. You need as many plants with the number of flowers for your class size. Alternatively, depending on your local environment, it may be possible for students to venture outside to select their own flowers. If you have a selection of plants available, allow students to do this investigation for two different plants. Direct students to compare the similarities and differences.

Equipment per pair of students: scalpel, hand lens, microscope, microscope slide, flower with petals attached.

Safety considerations

Students must take care when handling and using the scalpel. If students go outside, brief them on the areas where they cannot go and the possibility of dangerous plants or animals in the area.

Common errors to be aware of

Students will make few errors in removing the main parts but guidance may be needed for dissecting the carpel. Demonstrate how to hold the scalpel safely without risking injury to the fingers. Some students will get parts of the flower mixed up so allow reference to a large diagram on the board, in the textbook, or from their notes.

Supporting your students

Allow these students to work in a larger group that follows your lead. You should demonstrate how to carry out each step of the method for students to follow suit.

Challenging your students

Direct students to produce a guide to dissecting a flower. This should include safety precautions and a method. If students have access to a smartphone or a digital camera, this can be done electronically and photographs of their own dissection can be used.

Discussion points and scientific explanation

Use the different parts of the flower to lead a class discussion or quiz about the function of each part. Discuss the markings on the petal and how they lead insects to the pollen. The inside of the carpel can be discussed in detail and related to the students' knowledge of pollination and fertilisation.

Answers to workbook questions

1. Student drawings
2. Student shows formula (magnification = image size/actual size), working, and correct answer with units. The answer should be greater than 1.0 as the drawing should be larger than the petal.
3. Student correctly names parts of their flower with a brief outline of each function.

Practical investigation 16.2
Oxygen for germination

Planning the investigation

In this investigation, students will observe the demonstration of the effect of removing oxygen from germinating seeds. Students will evaluate the safety requirements for this investigation.

This investigation will focus on the following assessment objectives:

- AO3.1 Demonstrate knowledge of how to safely use techniques, apparatus and materials (including following a sequence of instructions where appropriate)
- AO3.3 Make and record observations, measurements and estimates
- AO3.4 Interpret and evaluate experimental observations and data.

Setting up for the investigation

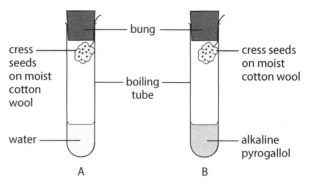

Figure 16.1

Set up the equipment as shown in Figure 16.1. Students may assist with the placing of cress seeds onto the cotton wool or the preparation of boiling tube A (the water control). Boiling tubes are suggested for this method but any glassware with a secure bung may be used, such as a conical flask or test-tube. Equipment required: approximately 25 ml of alkaline pryogallol, water, boiling tube × 2, bung for boiling tube × 2, cress seeds placed on moist cotton wool × 2, sewing thread, safety spectacles.

Safety considerations

Alkaline pryogallol is a caustic substance and must not be handled by students. Prepare boiling tube B in safe conditions prior to the investigation. Handling and storage should be referred to in your safety data sheets. Keep the alkaline pyrogallol in an airtight container at room temperature, away from direct sunlight. Wear safety spectacles and wash hands thoroughly after use. Do not allow any students to ingest or inhale the alkaline pyrogallol.

Common errors to be aware of

The bung may not be secure, allowing oxygen to get into the system. This would produce some growth but can be avoided with a secure bung being selected and tested beforehand.

Supporting your students

Some students may not be aware of what cress seeds grow into or what they look like. Either prepare some grown seeds in advance or display an image of them on the board. This will enable the students to visualise what the seeds are expected to do after growth.

Challenging your students

Direct students to research the safety information for alkaline pyrogallol and/or caustic substances.

Discussion points and scientific explanation

The alkaline pyrogallol absorbs the oxygen from the surroundings, preventing the seeds from germinating. If some air/oxygen is able to get into the boiling tube, then there may be minimal growth. Boiling tube A shows normal growth as all of the conditions for germination are there. Water is a control to show the effect that removing oxygen has on the seeds.

Answers to workbook questions

1. Student predictions (a–d)
2. Students describe the difference in growth – boiling tube A shows normal growth, boiling tube B shows little or no growth at all.
3. The removal of oxygen from the surroundings prevents germination as oxygen is required to do so.
4. As a control, to show that the removal of oxygen was a factor in the different results
5. Caustic substances are extremely harmful to the body and must be handled by a fully trained professional.
6. The seeds would begin to grow as the removal of oxygen would not have killed the seeds while in boiling tube B.

Practical investigation 16.3
Measuring the effect of temperature on the germination of cress seeds

Planning the investigation

In this investigation, students will plan their own investigation to investigate the effect of temperature on the germination of cress seeds.

This investigation will focus on the following assessment objectives:

- AO3.1 Demonstrate knowledge of how to safely use techniques, apparatus and materials (including following a sequence of instructions where appropriate)
- AO3.2 Plan experiments and investigations
- AO3.3 Make and record observations, measurements and estimates.

Setting up for the investigation

Purchase enough cress seeds for the class. Each group of students will require 15 seeds for the investigation as planned. Seek out a cold place where students can store their seeds, such as the departmental refrigerator. Students will need two suitable locations to store their other seeds for a medium and a higher temperature. Equipment per group of 2–4 students: cress seeds × 15, cotton wool or paper towels, Petri dish, access to areas of suitable temperatures.

Safety considerations

Students should wash their hands after contact with the seeds. The refrigerator should be safely accessible to the students; otherwise, a teacher or member of staff must place the seeds in the refrigerator.

Common errors to be aware of

Students may plan a method that does not differentiate the temperatures enough. Encourage students to use a refrigerated area, a medium temperature area such as room temperature, and an area that will have higher temperatures such as a warm cupboard, in direct sunlight, or above a heat source.

Supporting your students

Provide students with pre-determined areas for the different temperatures.

Challenging your students

Direct students to calculate the percentage difference between the different germination rates at different temperatures.

Discussion points and scientific explanation

Discuss how most plants have an optimum temperature for growth. Compare the results for the different temperatures; if extreme temperatures were used, then growth and germination should be much smaller than any seeds grown in the range of typical room temperature. The enzymes involved in successful plant growth will not work as effectively outside their optimum temperatures of between 20 °C and 30 °C.

Answers to workbook questions

1. Student method
2. Student table
3. Student calculations
4. Student uses data from their table to describe the differences in growth. Student explains that the cress seeds will grow best within the range of room temperature as that is when the enzymes involved in breaking down molecules that supply the seeds with the nutrients used for growth are most active.

Answers to exam-style questions

1.
   a Stamen [1]
   b Contains the filament/anther [1] and is the male reproductive organ of a flower [1]
   c Carpel [1]
   d Contains stigma/style/ovary [1] and is the female reproductive organ of a flower [1]
   e Student drawing larger than the petal in Figure 16.2 in the student workbook [1], labelled as the petal [1], smooth outline drawn in pencil [1], shape and size in proportion [1], drawn with sharp pencil [1]

17 Reproduction in humans

> **Overview**
>
> In this chapter, students will model how the human body prepares for giving birth to ensure that the fetus is not damaged by external forces.

Practical investigation 17.1
Protecting the fetus

Planning the investigation

In this investigation, students will observe how a model that represents the amniotic fluid in the uterus shows the function of the fluid in protecting the fetus from external damage. Obviously, the scope for practical investigation is limited in this unit but Practical investigation 17.1 will allow students to observe how the fetus is protected, using modelling. Scientists use modelling to help observe and visualise how things happen, especially when there are ethical and logistical barriers to observing the real events.

This investigation will focus on the following assessment objectives:
- AO3.1 Demonstrate knowledge of how to safely use techniques, apparatus and materials (including following a sequence of instructions where appropriate)
- AO3.3 Make and record observations, measurements and estimates.

Setting up for the investigation

Purchase enough eggs for student use. This can be carried out as a demonstration, in which case you can ensure success by hard-boiling the egg inside the water. If you do not have the space for dropping the egg from a height, the same effect can be shown by placing the egg inside a larger beaker of water. Shaking the beaker without water it will break the egg; shaking the egg in water will leave the egg intact. Line the testing area with newspapers to soak up water and/or egg that may escape from the bags.

Equipment per group of 2–6 students: eggs × 2, metre stick or measuring tape, sealable plastic bags × 2, newspapers.

Safety considerations

Clear a suitable area for testing, provide students with the means of covering their feet, and be wary of slippery surfaces if water or egg escapes onto the floor.

Common errors to be aware of

Students may not measure the 1 m height accurately; however, this can be avoided by using a reference point such as holding the egg-bag out horizontally before release. Students of different heights will be dropping from different heights but at least the height of the release is the same with and without the water in the bag.

Supporting your students

Students may observe other students carrying out the investigation and focus on recording what they can observe. Using prompt diagrams to show the egg in the bag of water next to a diagram of a fetus in the amniotic sac will guide students towards the connection that should be made.

Challenging your students

Direct students to prepare a 30-second speech about the importance of the amniotic sac and the amniotic fluid during pregnancy.

Discussion points and scientific explanation

The amniotic fluid is where the fetus floats freely in the uterus. This fluid protects the fetus from external forces and knocks. The water in the bag represents the amniotic fluid and was able to protect the egg from breaking as the force of hitting the ground was 'cushioned' by the surrounding water. Some discussion must take place to reaffirm the notion of modelling. Encourage students to discuss why it is not possible to observe a real fetus in the classroom or laboratory.

Answers to workbook questions

1. Egg in water: no breakage
   Egg without water: egg breaks
2. Amniotic fluid
3. Fetus
4. The fluid protects against external forces and knocks that the fetus may experience.
5. As a control/to compare what would happen if the water was not there to protect the egg

Answers to exam-style questions

1.
   a. Pituitary gland [1]
   b. The egg is released [1].
   c. Oestrogen [1] and progesterone [1]
   d. The egg has not been fertilised and progesterone ceases being secreted [1], so the uterus wall breaks down [1] and menstruation begins [1].
   e. The concentration of progesterone would not decrease [1] as progesterone will continue to be secreted until the fertilised egg [1] is fully sunken into the wall of the uterus [1].

2.
   a. Amniotic fluid [1]
   b. To support the embryo/fetus [1] and to protect it from external forces [1]
   c. The fetus has developed limbs [1] and the placenta is joined to the umbilical cord [1]
   d. Any three of the following: named nutrients/food materials, oxygen, carbon dioixde, waste materials [3]

18 Inheritance

Overview

In order to observe inheritance, it is necessary to observe what happens to different generations as genetic variation takes place from one generation to the next. Unless you remain at school for hundreds of years, this is not possible to investigate in your own laboratory. Scientists often use animals such as *Xenopus laevis* (African clawed frog) as these have a much shorter life cycle and genetic changes and development can be observed easily. As you do not have access to lots of African clawed frogs, this unit will look at how asexual reproduction produces clones. Students should apply their knowledge of sexual reproduction from the remainder of the unit to fully understand how genetic variation takes place in sexual reproduction. They will also use your knowledge of inheritance to predict, and observe, the ratios of producing offspring with a particular characteristic.

Practical investigation 18.1
Cloning a cauliflower

Planning the investigation

In order to observe inheritance, it is necessary to observe what happens to different generations. This is not possible in the school laboratory setting, although 'real' scientists will often use animals such as *Xenopus laevis* (African clawed frog) as these have a much shorter life cycle and genetic changes and development can be observed easily. It is important that the process of asexual reproduction is linked back to the genetic variation that one would expect to find if the students were able to observe sexual reproduction.

In this investigation, students will use their aseptic technique learned in Chapter 10 to clone a piece of cauliflower. This investigation will test your students' ability to follow a method closely and adhere to safety precautions.

This investigation will focus on the following assessment objectives:

- AO3.1 Demonstrate knowledge of how to safely use techniques, apparatus and materials (including following a sequence of instructions where appropriate)
- AO3.3 Make and record observations, measurements and estimates.

Setting up for the investigation

There are many methods to carry out this investigation and some specialist equipment, diluvials, and sterilants are available to ensure the success of the investigation by removing the risk of contamination. However, this method aims to provide you with the simplest method that can be used in most science laboratories with minimal expense or specialist solutions.

Prepare, or purchase, a basic agar growth solution for growth. Clean all glassware with hot, soapy water and double-rinse with distilled water before allowing to dry. Students or technicians to clean the work surface with a suitable alcohol or disinfectant solution.

Equipment list per group of 2–4 students: Floret of cauliflower, forceps, scalpel, disinfectant solution, 50 ml glass beaker, stopwatch, distilled water (sterile), 250 ml glass beaker × 2, petri dish × 2, Bunsen burner, test-tube containing agar growth medium × 2, aluminium foil, marker pen.

Safety considerations

Consult your safety data sheets and carry out a risk assessment prior to preparing this investigation. Students to wear gloves, safety spectacles, and lab coat or apron at all times. Take care when using or handling the Bunsen burner, forceps, and scalpel. Sterilise the work surface after use and all equipment used as recommended in your risk assessment and safety sheets. If the cauliflower was contaminated in any way, it will grow bacteria and fungi. If this happens, the contents of the test-tube must be suitably disposed of following the guidelines and the test-tube must be cleaned, preferably with an autoclave or in similar conditions.

Common errors to be aware of

Students may make mistakes following such a lengthy method with little margin for error. Strong reinforcement of the need to follow the method closely and the motivation of the possibility of cloning the cauliflower should keep the alertness levels high enough for students to be successful.

Some steps of the method must be carried out quickly to avoid contamination. Demonstration of how to do this as a team will assist students with avoiding contamination of the explants.

Supporting your students

Students will benefit from a diagram or flow chart on the board showing how each stage is achieved. Pairing students with a more able student may benefit as the more able student can take a lead role. If the method is too complicated then the students could try a similar method by replanting some cuttings from a plant. They can plan the method and observe the results after a similar time to the tissue culture of the cauliflower.

Challenging your students

Direct students to produce a cloning guide with a full set of instructions and diagrams to make the method easier to follow. This could then be used by less able students in the future.

Discussion points and scientific explanation

Revisit the process of mitosis and asexual reproduction. Discuss how this produces genetically identical clones to the explants. Discussion can lead to the advantages and disadvantages of cloning. Advantages include the ability to produce exact copies of a desired organism or feature, disadvantages include loss of diversity and potential for a species to be severely damaged by a sudden disaster or infection. The discussion should focus on how sexual reproduction does not produce clones and possibly lead into some ethical discussion about the following:

- Consider why students cannot observe sexual reproduction or genetic variation over time.
- Could cloning be used to reproduce 'desirable' human characteristics/genes?

Answers to workbook questions

1. Observations of cloned cauliflower:
   After one week: a few mm of growth
   After two weeks: a full floret should be visible
2. a DNA would be identical.
   b The cauliflower asexually reproduced by mitosis to produce new cells that are genetic copies of the original sample.
3. Contamination will lead to growth of bacteria and fungi, link to potential safety and health hazards if this happens. More importantly, the experiment will fail due to death of the plant if this was allowed to occur.

Answers to exam-style question

1. a Observable feature [1] of an organism
   b Has two identical alleles [1] of a particular gene [1]
   c i A version of a gene [1]
      ii An allele that is always expressed if present in the genotype
      iii An allele that is present in the genotype but is not expressed in the presence of a dominant allele [1]
   d Genotypes state as BB and bb [1], genetic cross drawn [1], possible genotypes shown [1], correct answer stated as 0% [1]
   e 100% [1]

19 Variation and natural selection

Overview

In this chapter, students will investigate how humans exhibit variation and the reasons behind this. They will investigate the adaptive features of a leaf that help those organisms to survive in their environment.

Practical investigation 19.1
Variation in humans

Planning the investigation

In this investigation, students will investigate some inherited characteristics within their class and use the data collected to present it as a tally chart, a histogram, and a bar chart.

This investigation will focus on the following assessment objectives:
- AO3.1 Demonstrate knowledge of how to safely use techniques, apparatus and materials (including following a sequence of instructions where appropriate)
- AO3.3 Make and record observations, measurements and estimates
- AO3.4 Interpret and evaluate experimental observations and data.

Setting up for the investigation

Very little set up is required, use of metre stick or tape measure is all that needs to be arranged.

Safety considerations

Take care when moving around the room as there will be many students moving in all directions..

Common errors to be aware of

Errors measuring height depending on the method used. Students may remove shoes to measure the true height but should put them back on before moving around the laboratory. Demonstrate to students how to accurately measure the height (head level and use of a ruler to measure from the top of the head to the scale.

Supporting your students

Direct students to work together in a group to support each other. Students can share their data rather than trying to rush to gather as much data as possible in the time permitted.

Challenging your students

Direct students to investigate other features such as tongue-rolling, gender, presence of earlobe creases, presence of freckles or any other characteristic that they may be interested in. Students may attempt to collect the original data for the entire class to improve reliability of their findings.

Discussion points and scientific explanation

Students should experience some variation for each of the characteristics investigated. Link this to the combination of DNA from mother and father, as well as potential mutations. Students should find a bell distribution curve for the hand size data, as most students will be around the average, with occasional extremes at either end of the range. Discuss which characteristics are the result of genetic and environmental factors. Some, such as mass, may be a combination of both. Encourage students to make a decision about whether a characteristic is genetic or environmental but to support their decisions with some reasoning.

> **Answers to workbook questions**
>
> 1. Student table
> 2. **a** Students correctly tally the number of students in each category. Categories should be evenly split across the range of data for that characteristic. Tally should use the correct tally method and have the final number in bracket next to the tallies.
>   **b** Histogram drawn to show data from tally chart (hand size and units on the x-axis, number of students on the y-axis), bars drawn touching each other in order of increasing or decreasing magnitude.
> 3. Bar chart drawn to show the number of students with each eye colour (eye colour on the x-axis, number of students on the y-axis), blocks should not be touching.
> 4. Student describes the pattern for the data of their data/chart/histogram.
> 5. Continuous
> 6. Height, hand size, eye colour, hair colour (accept if mass is not included)
> 7. Increase the sample size and survey more students, i.e. the idea that the more students that are surveyed, the more reliable the data will be.

Practical investigation 19.2 Adaptive features

Planning the investigation

In this investigation, students will investigate stomata, an adaptive feature of most plant leaves designed to ensure their growth and survival in their environment. There is little variation between plants in the same area but plants will have a different number of stomata depending on their environmental conditions. This investigation aims to make the link between plants that have many stoma and the amount of carbon dioxide that they need for photosynthesis.

This investigation will focus on the following assessment objectives:

- AO3.1 Demonstrate knowledge of how to safely use techniques, apparatus and materials (including following a sequence of instructions where appropriate)
- AO3.3 Make and record observations, measurements and estimates
- AO3.5 Evaluate methods and suggest possible improvements.

Setting up for the investigation

Purchase enough plants to provide each group of 2-4 students with two leaves each. Prior to the investigation, place one plant in darkness for 24 hours and the other plant in normal, sunny conditions. The plants should be the same species. Any plant with a leaf is sufficient for this investigation.

Equipment per group of 2-4 students: leaf from plant in darkness, leaf from plant in sunlight, microscope slide, light microscope, clear nail varnish, transparent cellophane tape.

It is possible to do this without the cellophane tape by using a scalpel to carefully lift the clear nail varnish from the leaf. This is slightly more difficult than the tape method and includes the slight increase of risk to safety by using a sharp scalpel.

Safety considerations

MH

- Take care if handling a scalpel.
- Wash hands afterwards if students come into contact with nail varnish.

Common errors to be aware of

Students may not apply a thick enough coat of varnish or they may try to lift the varnish while it is still not dry. If no stomata are found in the section applied, repeat on a different leaf or on a different area of the same leaf.

Supporting your students

Walk around the room to help students who may be struggling with the lifting of the varnish. Students who do this quickly can be assigned as helpers to those who are struggling.

Challenging your students

Direct students to plan the method for direct comparison of the same leaf from the same plant under the two different conditions of sunlight and darkness. Permit these students to carry out the investigation if time is available.

Discussion points and scientific explanation

The stomata of most species of plant will open in sunlight and close in darkness. This should be reflected in the number of open stomata in the investigation. If not, suggest experimental error as there may be different numbers of stomata in the different areas of the different leaves. The stomata of some plants close during the day to minimise loss of water by evapotranspiration. Whether a plant has many stoma or not, may be linked to that particular plant, in that particular environment, adapting to its own needs. This may mean some plants having fewer open stomata to reduce water loss in hotter climates, or plants that may have more stomata as they require more carbon dioxide to do more photosynthesis in order to meet their energy needs through respiration. Encourage students to discuss why plants with many more/less stoma may not pass on their characteristics to the next generation.

Answers to workbook questions

1. Student accurately counts the number of stomata shown for each leaf.
2. Student describes the difference in open stomata and links this to the need for the plant to prevent water loss and/or to take in more carbon dioxide.
3. More reliable test, to compare the effect of the amount sunlight
4. Use the same leaf, place in sunlight and take a varnish sample. Then, place the same leaf into darkness and repeat the observation. The leaf must not be removed from the plant and it must be fed and watered as normal.

Answers to exam-style question

1.
   a. Continuous
   b. Histogram drawn [1], blocks touching [1], blocks drawn in increasing or decreasing magnitude [1], suitable axis labelled [1]
   c. Similar pattern described for boys, girls, or both, of few students at the extreme [1] and most students gathered around the mean [1]
   d. Large sample of the population gathered [1]

20 Organisms and their environment

Overview

In this chapter, students will investigate how organisms interact with their environment and investigate how population sizes may affect the biotic factors around them.

Practical investigation 20.1 Using a quadrat

Planning the investigation

In this investigation, students will carry out random quadrat sampling to investigate the population sizes of different species in a local area. Students will link the population sizes to factors such as sunlight available and predators in the area.

This investigation will focus on the following assessment objectives:

- AO3.1 Demonstrate knowledge of how to safely use techniques, apparatus and materials (including following a sequence of instructions where appropriate)
- AO3.3 Make and record observations, measurements and estimates
- AO3.4 Interpret and evaluate experimental observations and data.

Setting up for the investigation

You will need to survey the type of school and the local community before carrying out this investigation. Ideally, a large field or desert area with some plant life will be available within your school grounds. Ensure that you have permission from your management team to use the area planned and check with other departments such as Physical Education that they are not already in use.

Equipment per group of 2-4 students: quadrat and smartphone or digital camera.

Not all students will have smartphones; group them accordingly so that each group has access to a smartphone. If no students have smartphones, they can make drawings or take notes to research their unknown species. Quadrats can be bought but if this is not suitable for your school, a homemade quadrat can be fashioned using metre sticks. If these are too flimsy to be thrown, some bias will have to be introduced and allow students to use the metre sticks (or string) to plot their quadrats without throwing.

If you have access to a light intensity sensor, introduce this to the students for comparing the light intensity of different areas being sampled. This will provide extra data for students, particularly the more able, to compare.

Safety considerations

Do not allow students to remove organisms or touch them with their hands, especially if they do not know what they are touching or observing.

Common errors to be aware of

Students will sometimes struggle to find enough organisms in their provided area. Your planning is key to securing an area most likely to give a range of results, ideally with some variation due to environmental factors. Students may count incorrectly; eliminate this problem by encouraging them to count more than once and take the mean number counted.

Supporting your students

Work closely with these students and demonstrate how to use the quadrat for the first sample or two. Once you are confident that students are sure of how to use the quadrat, you should be able to leave them to investigate independently. It may be beneficial for you to identify in advance the names of different species to focus on. This will allow data to be shared between groups.

Challenging your students

Direct students to write a brief article about the diversity of the population for the school newsletter or website. Students can make recommendations for change if a particular area requires more sunlight or water.

Discussion points and scientific explanation

Discuss why quadrats are used - it would not be possible to count every organism for every species. Random sampling, if enough samples are carried out, can be used to make an accurate prediction of the population size for a particular organism. Some areas that show lower population sizes can be linked to their environmental factors, such as lack of sunlight or water. It may also be an area that is traveled on by humans or animals.

Answers to workbook questions

1. Student table
2. Student bar chart
3. Student describes the pattern of data gathered.
4. Student makes comparison between two different areas that have different population sizes for the same species.
5. Student refers to factors such as light, water, or available food. Student may also refer to factors such as human footfall or the presence of predators.
6. To improve the reliability of the data taken
7. This would take too long and there may be change over the time taken anyway.
8. To reduce bias and improve validity of the method

Practical investigation 20.2
Making compost

Planning the investigation

In this investigation, students will make their own compost using different types of waste. Students will observe their compost for several weeks and link the decomposition of organic waste to their knowledge of decomposition and the nitrogen cycle.

This investigation will focus on the following assessment objectives:

- AO3.1 Demonstrate knowledge of how to safely use techniques, apparatus and materials (including following a sequence of instructions where appropriate)
- AO3.3 Make and record observations, measurements and estimates.

Setting up for the investigation

There are two choices for the collection of waste. You can either organise this yourself, using household or school waste, or you can direct students to bring in waste from a predetermined list. It may be easier and safer for you to organise your own waste, especially if you have links with the school cafeteria or other food services.

Examples of the types of waste required are:

- (Organic) Kitchen waste: vegetable and fruit peelings, coffee grounds, eggshells, food scraps. Do not use dairy or meat.
- Organic waste: sawdust, grass clippings, straw, leaves, leaf litter, paper.
- Non-biodegradable waste: aluminium foil, plastic, Styrofoam, polystyrene.

Equipment per group of 2–5 students: 2-litre plastic bottle, scalpel or knife, 1 kg soil (this can be reduced to 0.5 kg for larger classes), organic kitchen waste, organic garden waste, non-biodegradable waste, mounted needle, Bunsen burner, water, pipette, aluminium foil, elastic band.

If you have access to worms, these can be added to speed up the decomposition process. Keep the compost out of direct light if you use worms.

Safety considerations
MH

Take care when using a sharp scalpel, cover jagged edges of plastic to avoid injury, wash hands (or wear gloves) after handling waste and soil. Do not add dairy or meat products to the compost.

Use of the heated mounted needle for making holes in the plastic bottle needs consideration.

Common errors to be aware of

Students may pack their layers too tightly. This should be avoided to allow enough air to circulate around the compost column. Gently shaking the compost occasionally should loosen the contents enough to allow the circulation of air.

Supporting your students

Students observe you preparing the compost and simply observe which items degrade and which items do not.

Challenging your students

Direct students to make predictions about each item of waste.

Discussion points and scientific explanation

Discuss the items that did decay quickly, such as the organic waste. Discuss the contribution to the nitrogen cycle that some foods might make. Discuss how a plant would take up the recycled nitrates and nitrites to keep the cycle going. Discuss how many years it would take for some items to be broken down and why we should avoid using these types of items.

Answers to workbook questions

1. Four-week date and observation: soil drying out, food breaking down, some evidence of mould
   - Eight-week date and observation: food broken down, biodegradable waste starting to break down
   - Twelve-week date and observation: all/most of food/kitchen waste decomposed completely and most biodegradable waste broken down. Non-biodegradable waste still evident.
2. Organic waste decomposes quickly; other waste will not decompose in the time allowed.
3. The layer that contains the organic kitchen waste
4. Decomposition
5. The rate would increase as the worms help to decompose some types of waste.

Answers to exam-style question

1.
   a Lag phase [1]
   b Log or exponential phase [1]
   c Stationary phase [1]
   d Death phase [1]
   e The conditions for growth are ideal [1], large number of cells reproducing quickly at the same time [1].
   f Student suggests lack of food or reduction in the ideal conditions [1], so some cells begin to die [1], the death rate exceeds the birth rate [1].

21 Biotechnology

Overview

In this chapter, students will investigate how bacteria are used in biotechnology and in the production of food and drink. They will observe and investigate the use of pectinase in industry and at home.

Practical investigation 21.1
Effect of pectinase on apple pulp

Planning the investigation

In this investigation, students will investigate the effect of pectinase on the production of juice from apple pulp. This investigation will focus on the following assessment objectives:

- AO3.1 Demonstrate knowledge of how to safely use techniques, apparatus and materials (including following a sequence of instructions where appropriate)
- AO3.3 Make and record observations, measurements and estimates
- AO3.4 Interpret and evaluate experimental observations and data.

Setting up for the investigation

Advance purchase of enough apples for the class is required. Each group will require 100 g of apple but it is best to buy a few extra apples to account for mistakes and peeling. Pectinase can be purchased from most science suppliers in advance.
Equipment list per group of 2–4 students: 100 g of apple, 250 ml beakers × 2, 100 cm³ measuring cylinder × 2, glass rods × 2, stopwatch, balance, 2 ml pectinase, safety spectacles, water bath or incubator, knife, chopping board, Clingfilm, funnel × 2, filter paper × 2.
A water bath can be made using a large beaker or polystyrene box/container with water at 40 °C inside.

Safety considerations

Students wear safety spectacles; take care when handling knives and the pectinase; wash hands immediately if pectinase comes into contact with skin. Refer to safety data sheets for safe storage and handling of pectinase.

Common errors to be aware of

Students may cut the apple pieces too large; circulate the room and check their chopping before allowing them to add pectinase and water. If a home-made water bath is used, the temperature will need to be monitored and kept as close to 40 °C as possible.

Supporting your students

Assist students with accurate reading of the measuring cylinder. This skill can be practiced during the incubation period by adding water to a measuring cylinder and observing the different volumes.

Challenging your students

Direct students to calculate the percentage difference in the amount of juice produced between the two measuring cylinders.

Discussion points and scientific explanation

Discuss how the pectinase catalysed the breakdown of pectin in the cell walls. Relate this to knowledge of enzymes as biological catalysts, and the knowledge of the presence of pectin in cell walls. The pulp was softened and more juice was released by the apple that had pectinase present in the beaker. Discuss the optimum temperature for enzymes such as pectinase to work in (37 °C), hence the use of the water bath or incubator at 40 °C.

Answers to workbook questions

1 Student results in table
2 Student graph
3 The pectinase caused more juice to be released from the apple than the apple that was in water only.
4 Pectinase broke down the pectin in the cell walls, releasing the juice more easily and faster.
5 As a control, to compare the effect of pectinase, to confirm that it was pectinase that caused the release of the juice from the pulp

Practical investigation 21.2
Effect of temperature on pectinase

Planning the investigation

In this investigation, students will plan their own investigation into the effect of temperature on pectinase, using the knowledge and method from Practical investigation 21.1.
This investigation will focus on the following assessment objectives:
- AO3.1 Demonstrate knowledge of how to safely use techniques, apparatus and materials (including following a sequence of instructions where appropriate)
- AO3.2 Plan experiments and investigations
- AO3.3 Make and record observations, measurements and estimates
- AO3.4 Interpret and evaluate experimental observations and data.

Setting up for the investigation

Advance purchase of enough apples for the class is required. Each group will require at least 100 g of apples but it is best to buy a few extra apples to account for mistakes and peeling. You may also wish to vary the investigation by using different fruits that you have available. Pectinase can be purchased from most science suppliers in advance.

Equipment list per group of 2–4 students: 100 g of apple, 250 ml beakers, 100 cm^3 measuring cylinders, glass rods, stopwatch, balance, pectinase, safety spectacles, water bath or incubator, knife, chopping board, Clingfilm, funnels, filter paper. Each group will need three of each item from the list; if you do not have the resources then different groups can investigate the different temperatures and collate their results.

A water bath can be made using a large beaker or polystyrene box/container with water at 40 °C inside. It is recommended that you use three very different temperatures for the water baths, such as 0 °C, 30 °C, and 50 °C in order to get clear results for comparison.

Safety considerations

MH

Students to wear safety spectacles; take care when handling knives and the pectinase, wash hands immediately if pectinase comes into contact with skin/eyes. Refer to safety data sheets for safe storage and handling of pectinase.

Common errors to be aware of

Remind students of the importance of cutting the fruit into small pieces and too check their efforts before they add the pectinase. If a water bath is used, the temperature will need to be monitored and kept as close to the agreed temperatures as possible.

Supporting your students

Direct students to carry out the method from Practical investigation 21.1 but only allow student to test two different temperatures. Test at 30 °C, and 50 °C to obtain clear results that are simpler to analyse and understand.

Challenging your students

Direct students to research and discuss how enzymes such as pectinase are now used to peel citrus fruits for commercial use.

Discussion points and scientific explanation

Discuss the optimum temperature for pectinase, which should be around 35–40 °C. Recall that enzymes work best at this temperature, i.e. the amount of pectin broken down increases and the amount of juice produced is greater.

> **Answers to workbook questions**
>
> 1. Student method
> 2. Student temperatures
> 3. Two safety precautions
> 4. Student table
> 5. Student answer of the best temperature
> 6. Produced the most juice
> 7. This is close to the optimum temperature for most enzymes. The enzyme will carry out more reactions at this temperature and in the case of pectinase; will break down more pectin and release more juice from the fruit.
> 8. Enzymes can be used to maximise the yield of juice produced from fruit on a commercial level.
> 9. Repeat the investigation and take a mean for the volume of juice produced; ensure that all variables are constant.

Practical investigation 21.3 Biological washing powders

Planning the investigation

In this investigation, students will investigate the effect of biological and non-biological washing powders on the removal of egg stains.

This investigation will focus on the following assessment objectives:

- AO3.1 Demonstrate knowledge of how to safely use techniques, apparatus and materials (including following a sequence of instructions where appropriate)
- AO3.2 Plan experiments and investigations
- AO3.3 Make and record observations, measurements and estimates
- AO3.4 Interpret and evaluate experimental observations and data.

Setting up for the investigation

Advance purchase of enough eggs for the number of students in the class. The white material can be made from old tee shirts, an old bedsheet, tea towels, or other similar items that can be cut up and divided among the groups. Purchase the two different washing powders. Equipment list per group of 2–4 students: white material × 2, egg, paper towels, spoon, plastic tray, 250 ml glass beaker × 2, 100 ml glass beaker, stirring rod, kettle, biological washing powder, non-biological washing powder.

Safety considerations

Wash hands and surfaces at the end of the investigation. Wash and reuse the white material for future investigations.

Common errors to be aware of

The egg can be washed out at a similar rate by both powders if left to soak in the solution for too long. Either test beforehand, or observe your demonstration closely and direct students to alter the wait time accordingly. Students may get confused, or forget, about which beaker is which one. Label the beakers to be very clear about what is inside each beaker. The material may not dry the material in the time provided in the lesson. Allow students to dry the material outside if the weather is suitable, on a radiator if available, with a hairdryer if available, or to be labeled and revisited in the next lesson.

Supporting your students

Prepare the eggy solution and prepare washing powder beakers in advance. Advance staining with the egg can benefit too, as students then focus on the results rather than the process.

Challenging your students

Direct students to design an advertisement for the use of biological washing powders. Their advert should aim to convince the reader why biological washing powders are better and explain the science behind this.

Discussion points and scientific explanation

The enzymes in the biological washing powder break down the fats and proteins that make up the stain, allowing them to wash out from the material faster than in the non-biological washing powder. Discuss how modern washing powders can work at high temperatures as special enzymes are used that have a much higher optimum temperature. This enables us to get rid of more dirt, grease, and stains from our clothes.

Answers to workbook questions

1. Student drawings
2. Student describes the difference between the stains drawn in their notes.
3. The biological washing powder contains enzymes that break down the stains faster so that more stain was removed compared to the non-biological washing powder.
4. Protease, accept other option used in washing powders, such as lipase
5. Student method suggests washing materials with similar stains at different temperatures for comparison. Some effort made to identify and control the variables, safety points should be noted.

Answers to exam-style questions

1. a Award one mark for any of the following points. [Maximum 6]
   - Named use of pectinase, such as commercial fruit juice, skin peeling
   - Protease, lipase used in washing powders, detergents, baby food
   - Enzymes used to catalyse reactions involved
   - Faster reactions
   - Lower temperatures required
   - Higher yield
   - Any other sensible point related to common use of the enzymes

   b Award one mark for any of the following points. [Maximum 5]
   - Suitable equipment identified
   - Fruit named
   - Fruit chopped into small pieces and pectinase added
   - Repeated at different pH levels for the enzyme
   - Enzyme and fruit pulp incubated at optimum temperature
   - Fruit pulp filtered into measuring cylinder
   - Comparison of volume of fruit juice collected
   - Safety points mentioned
   - Repeat testing for reliability

22 Humans and the environment

Overview

In this chapter, students will investigate how humans have affected the environment and how this can be measured. They will use this information to make assumptions about what has caused the changes and what needs to happen in order to prevent further damage.

Practical investigation 22.1
Effect of acid on the germination of cress seeds

Planning the investigation

In this investigation, students will investigate and observe how acid 'rain' affects the germination of cress seeds. This investigation will focus on the following assessment objectives:
- AO3.1 Demonstrate knowledge of how to safely use techniques, apparatus and materials (including following a sequence of instructions where appropriate)
- AO3.3 Make and record observations, measurements and estimates
- AO3.5 Evaluate methods and suggest possible improvements.

Setting up for the investigation

Purchase cress seeds in advance; you will need enough seeds to give each group approximately 40 seeds. You can use as few as one seed per petri dish/concentration but between 5 and 8 provides a reasonable sample for comparison.
Prepare some different concentrations of acid, such as 0.001 M, 0.01 M, 0.1 M, 1.0 M or 0.1 M, 0.2 M, 0.4 M, 0.8 M. Any laboratory acid can be used for this, such as hydrochloric acid or sulfuric acid.
Equipment list per group of 24 students: cress seeds, acids of different strength, pipettes, Petri dishes, paper towels, water.

Safety considerations

Wear safety spectacles at all times. Wash hands immediately if in contact with the acid.

Common errors to be aware of

Do not over-water the seeds; just provide enough to dampen the paper towel that they are on. You may need to 'water' them during the seven days; this can be done yourself or by the students depending on the days they have their lessons with you.

Supporting your students

Provide only one concentration of acid for direct comparison to the water. The difference in growth will be more pronounced and easier to observe.

Challenging your students

Direct students to measure the growth of the seeds and investigate if there is a relationship between the relative strength of the acid used and the height grown by the cress.

Discussion points and scientific explanation

Link the acid used to acid rain and discuss how acid rain has an adverse effect on the growth of crops and plants in that area. Relate to the causes of acid rain and the type of people that might be affected by poor crop growth. This may include farmers or local communities that rely on the crops for food.

> **Answers to workbook questions**
>
> 1. Completed student table
> 2. Student describes the pattern of data from their investigation/table. Students link this to the effect of acid rain on soil pH and that the optimum pH for most plant/crop growth is around pH 7.
> 3. The acid used represents acid rain and the growth of the cress seeds shows what might happen to crops/plant life if acid rain falls and reduces the pH of the soil.
> 4. The height/mass of the cress could be measured to compare the growth under different acidic conditions.

Practical investigation 22.2 Fossil fuel combustion

Planning the investigation

In this investigation, students will investigate and observe what happens when coal is heated. Students will observe how the gases formed are potentially harmful to the environment.

This investigation will focus on the following assessment objectives:
- AO3.1 Demonstrate knowledge of how to safely use techniques, apparatus and materials (including following a sequence of instructions where appropriate)
- AO3.3 Make and record observations, measurements and estimates
- AO3.4 Interpret and evaluate experimental observations and data
- AO3.5 Evaluate methods and suggest possible improvements.

Setting up for the investigation

Set up the apparatus in advance of the lesson for demonstration to the students. This is safer to carry out in a fume cupboard as a demonstration.

Your equipment needs may be different from what Figure 22.1 suggests. The equipment can be modified by using heat-resistant boiling tubes with a delivery system into a conical flask or boiling tubes containing only the cotton wool or the Universal Indicator. If you choose to do only one of these, select the Universal Indicator as this will show the effect of the gases produced on the pH of the solution. Equipment list for class demonstration: equipment as shown in Figure 22.1.

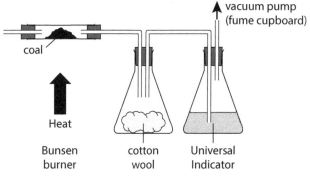

Figure 22.1

Safety considerations

Carry out the investigation in a fume cupboard or with a suitable vacuum pump to deal with the gases produced. Do not handle hot equipment after heating with the Bunsen burner.

Supporting your students

Before carrying out the demonstration, remind students of the pH scale and demonstrate the colour changes in acidic and alkaline conditions.

Challenging your students

Allow students to set up a second investigation, using the limewater and observe the production of carbon dioxide when burning the coal. This should be done under your supervision but the students can plan and prepare the equipment ready for you to heat the coal.

Discussion points and scientific explanation

The burning of coals releases smoke particles, which turns the cotton wool dark/brown/black. The burning also produces gases such as sulfur dioxide and carbon dioxide, which are acidic. This is shown as the gas is bubbled into the Universal Indicator solution, the solution turns orange/red. This shows the presence of acidic gases; show a pH scale on the board to reinforce this.

Answers to workbook questions

1. Student diagram
2. **a** Cotton wool turned brown/black.
   b Universal Indicator turned orange/pink/red.
3. Shows that the gases produced are acidic. These gases dissolve in water vapour and condense in the air as clouds before falling as acidic rain.
4. Limewater turns cloudy as the burning of coal produces carbon dioxide. Limewater turning cloudy is a positive test for carbon dioxide.
5. Student outlines an investigation that burns different quantities of coal. The coal is heated until completion and the pH of the gases produced are measured using a pH probe or a pH scale.

Answers to exam-style question

1. One mark for any point from below.
   - Named fossil fuels burnt
   - Fossil fuels contain sulfur.
   - Burning fossil fuels produces sulfur dioxide and nitrogen oxides.
   - Gases dissolved in water vapour.
   - Acid rain falls.
   - Soils become more acidic.
   - Plant growth is inhibited by the acidic soil.
   - Trees are damaged by sulfur dioxide.
   - Acid rain falls into rivers and waterways.
   - Acid rain in rivers adds toxic chemicals and reduces fish life due to food shortage.